CONTENTS

Chapter 1
Movements in the Info-communications Industry as a Whole

Chapter 2
Situation of Info-communications Service Usage

Chapter 3
Situation of TCA Members

Chapter 1
Movements in the Info-communications Industry as a Whole

1-1 Trends in Number of Telecommunications Carriers

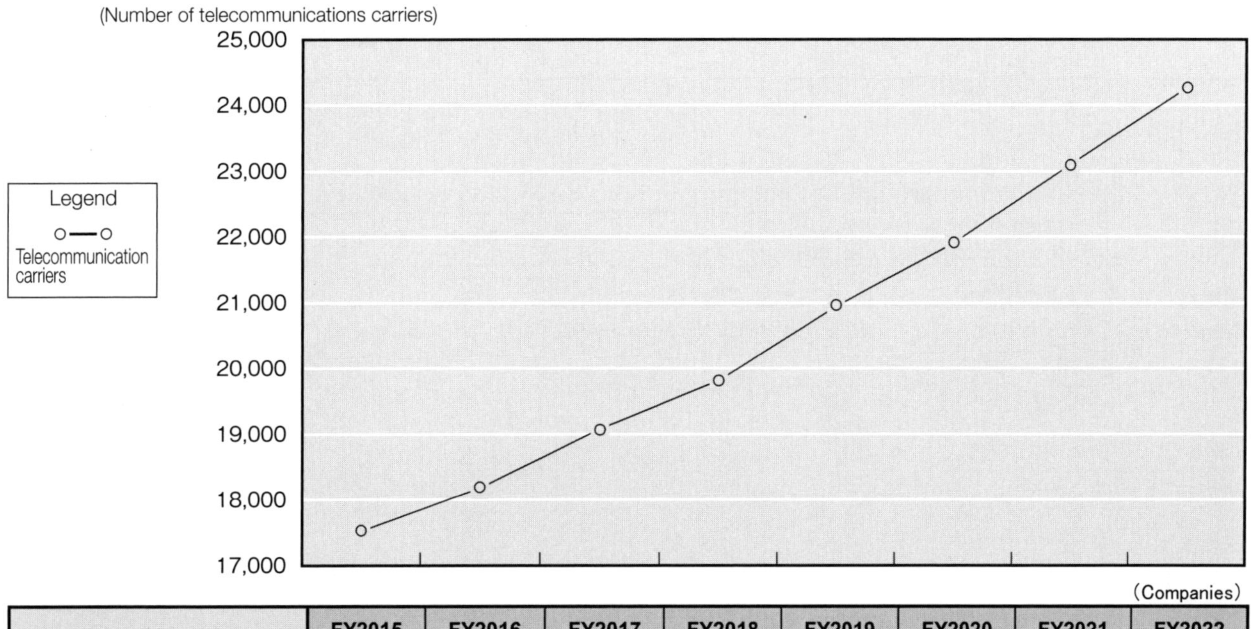

(Companies)

	FY2015	FY2016	FY2017	FY2018	FY2019	FY2020	FY2021	FY2022
Number of telecommunications	17,519	18,177	19,079	19,818	20,947	21,913	23,111	24,272

*Compiled by TCA based on data publicized by the Ministry of Internal Affairs and Communications

1-2 Telecommunications and Broadcasting Industries Sales in Their Relevant Business Activities

(Million yen)

		Telecommunications industry	Broadcasting industry		
				Commercial broadcasting	Cable television broadcasting
FY2021	Sales in relevant business activities	14,725,592	2,861,380	2,369,738	491,643

*Compiled by TCA based on data publicized by the Ministry of Internal Affairs and Communications

1-3 Amounts of Investment in Plant and Equipment Acquired by the Telecommunications and Broadcasting Industries

(Million yen)

			Telecommunications industry	Broadcasting industry	Commercial broadcasting	Cable television broadcasting
FY2021	Amounts of investment in plant and equipment		3,261,113	226,929	97,838	129,091
		Amounts of investment in plant and equipment (except software)	2,695,798	213,259	84,696	128,563
		Software	565,315	13,670	13,142	528

*Compiled by TCA based on data publicized by the Ministry of Internal Affairs and Communications

1-4 Number of Employees by Type of Employment in the Telecommunications and Broadcasting Industries

(Number of employees)

				Telecommunications industry	Broadcasting industry	Commercial broadcasting	Cable television broadcasting
FY2021	Number of employees			120,154	35,021	25,661	9,359
		Number of regular employees		120,103	34,672	25,322	9,351
			Permanent employees (excluding employees on loan to other companies/ organizations)	89,638	28,503	20,301	8,202
			Non-permanent employees (part-time workers, etc.)	13,551	3,760	3,157	603
			Employees detached to other companies/ organizations	11,478	963	953	10
		Temporary employees		51	348	340	9
	Dispatched employees			28,798	7,306	6,223	1,083

*Compiled by TCA based on data publicized by the Ministry of Internal Affairs and Communications

Chapter 2
Situation of Info-communications Service Usage

2-1 Situation of Number of Contracts for Various Services

2-1-1 Trends in Number of Telecommunications Services Subscriptions, etc.

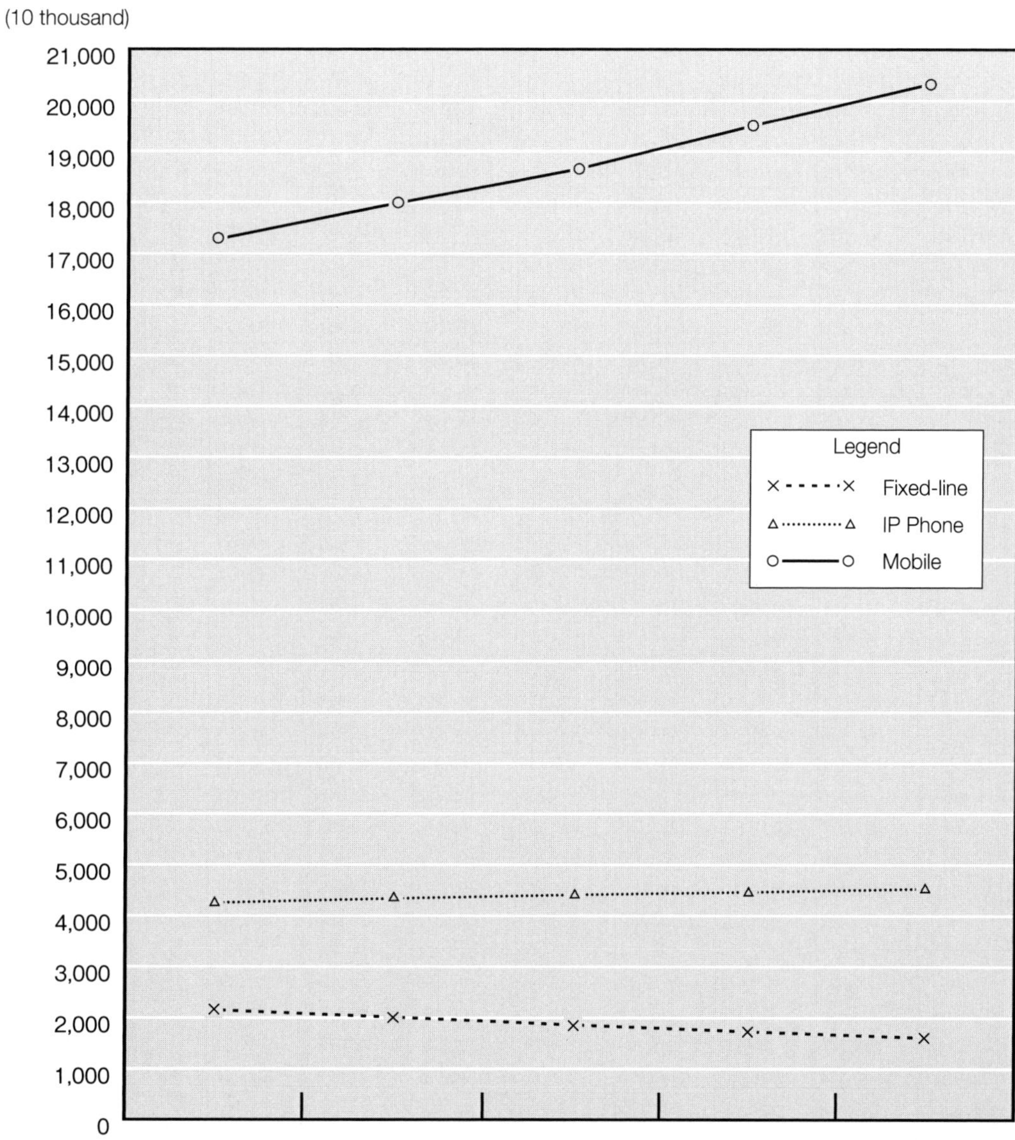

(Units: 10,000 contracts (subscriber telephones, ISDN, mobile phones, and PHS); 10,000 units (public phones); and 10,000 telephone numbers (IP phones))

Service		FY2017	FY2018	FY2019	FY2020	FY2021
Fixed-line Service Total		2,151	2,011	1,861	1,731	1,608
	Subscriber Telephone	1,845	1,724	1,595	1,486	1,383
	ISDN	290	272	251	231	212
	Public Phone	16	16	15	15	14
IP Phone		4,255	4,341	4,413	4,467	4,535
	(0ABJ-IP phone)	3,364	3,446	3,521	3,568	3,594
	(050-IP phone)	891	895	892	899	941
Mobile Service Total		17,279	17,987	18,651	19,505	20,333
	Mobile Phone	17,019	17,782	18,490	19,440	20,300
	PHS	260	206	162	66	34

Note: Figures for "Public Phone" represent the numbers of installed units.
*Compiled by TCA based on data publicized by the Ministry of Internal Affairs and Communications

2-1-2 Trends in Number of Subscriber Telephone Contracts by Prefecture

(Contracts)

Pref.	FY2018	FY2019	FY2020	FY2021			
					NTT(Re-entry)		
	Total	Total	Total	Total	Total	Business	Residential
Hokkaido	923,739	851,620	796,415	743,717	718,592	130,070	588,522
Aomori	251,263	232,337	220,235	207,836	201,465	33,884	167,581
Iwate	233,019	216,909	206,255	195,555	189,134	31,841	157,293
Miyagi	318,343	296,178	279,251	262,285	249,300	51,947	197,353
Akita	188,956	176,344	167,366	157,805	152,694	26,374	126,320
Yamagata	160,955	149,321	140,238	132,005	127,450	23,946	103,504
Fukushima	307,809	285,623	269,270	253,791	246,796	44,962	201,834
Ibaraki	400,105	370,700	348,577	327,492	315,412	58,413	256,999
Tochigi	266,751	247,955	232,351	217,990	209,306	39,219	170,087
Gunma	276,539	258,205	242,358	227,896	219,360	38,469	180,891
Saitama	817,897	757,130	708,569	662,367	628,290	112,982	515,308
Chiba	715,804	663,591	621,850	582,594	552,673	108,800	443,873
Tokyo	1,876,185	1,746,802	1,632,327	1,520,096	1,390,606	454,025	936,581
Kanagawa	1,041,101	962,496	895,725	834,460	781,230	169,041	612,189
Niigata	335,803	311,268	291,960	274,152	263,551	52,220	211,331
Toyama	139,585	127,722	117,353	107,889	103,384	24,388	78,996
Ishikawa	159,298	149,183	140,245	130,953	125,413	27,820	97,593
Fukui	88,915	81,638	75,826	70,586	67,957	19,282	48,675
Yamanashi	134,501	123,877	115,143	107,101	103,915	21,639	82,276
Nagano	324,681	297,636	275,624	256,273	246,752	54,469	192,283
Gifu	265,742	245,433	227,804	210,970	202,352	48,209	154,143
Shizuoka	494,447	454,097	416,951	385,348	364,154	84,529	279,625
Aichi	809,403	745,776	690,630	637,741	596,793	155,033	441,760
Mie	245,304	223,625	205,111	187,747	180,948	42,516	138,432
Shiga	138,045	128,055	119,017	111,205	105,898	28,047	77,851
Kyoto	344,377	319,745	297,333	276,557	260,618	59,859	200,759
Osaka	1,093,866	1,007,276	933,172	865,163	793,334	218,116	575,218
Hyogo	559,365	518,001	481,673	447,877	422,227	106,946	315,281
Nara	164,482	152,252	140,712	130,292	123,070	25,381	97,689
Wakayama	148,574	137,894	128,224	120,004	116,267	24,344	91,923
Tottori	81,943	76,073	71,072	66,590	64,595	15,068	49,527
Shimane	134,306	125,435	115,811	108,523	106,465	21,577	84,888
Okayama	286,727	266,902	248,164	230,916	221,500	45,270	176,230
Hiroshima	446,484	416,457	389,825	364,071	349,054	71,100	277,954
Yamaguchi	272,802	254,499	237,910	222,252	217,206	35,038	182,168
Tokushima	113,946	104,816	96,540	89,760	86,980	19,867	67,113
Kagawa	139,600	128,440	118,793	109,843	103,959	22,997	80,962
Ehime	234,922	217,179	201,157	185,154	179,897	34,474	145,423
Kochi	141,651	130,410	121,011	112,224	109,592	22,213	87,379
Fukuoka	661,901	608,481	561,601	516,562	484,762	108,627	376,135
Saga	109,016	100,260	92,939	85,521	82,462	16,835	65,627
Nagasaki	256,654	237,908	220,404	203,293	197,239	36,803	160,436
Kumamoto	280,380	260,663	240,309	222,847	216,414	40,246	176,168
Oita	203,951	188,985	175,422	162,783	157,569	30,667	126,902
Miyazaki	175,738	160,800	148,004	135,501	131,637	24,596	107,041
Kagoshima	315,219	290,522	264,769	241,866	235,855	41,897	193,958
Okinawa	162,126	147,354	134,449	123,643	119,111	29,573	89,538
Total	**17,242,220**	**15,953,873**	**14,855,745**	**13,827,096**	**13,123,238**	**2,933,619**	**10,189,619**

*Compiled by TCA based on data publicized by the Ministry of Internal Affairs and Communications and other organizations

NTT Subscribers by Prefecture (FY 2021)

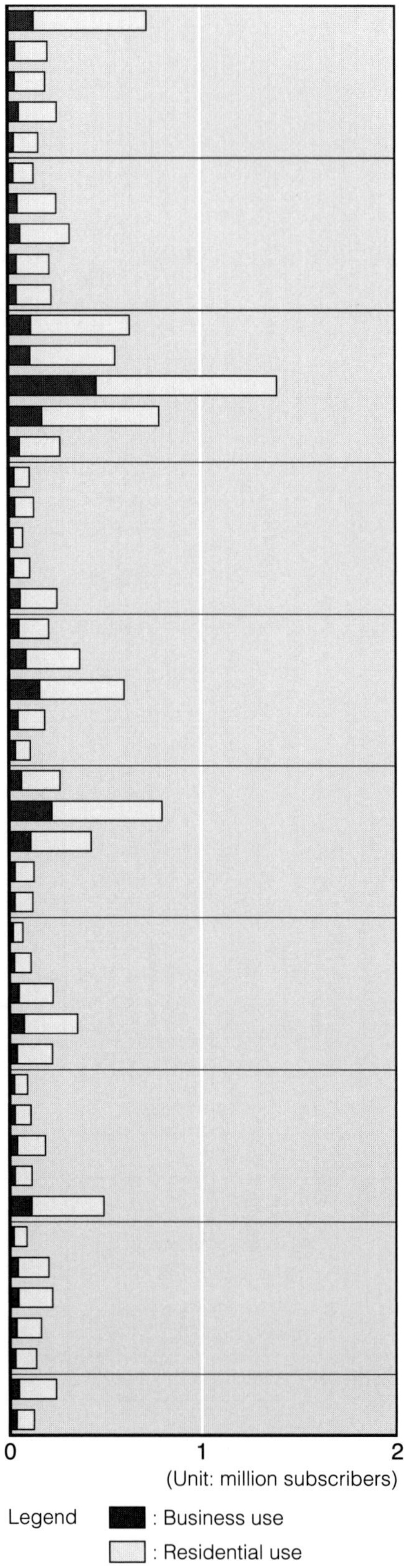

0 1 2

(Unit: million subscribers)

Legend ■ : Business use □ : Residential use

2-1-3 Trends in Number of ISDN Contracts by Prefecture

(Contracts)

Pref.	Basic Interface							Primary Rate Interface				
	FY2018	FY2019	FY2020	FY2021				FY2018	FY2019	FY2020	FY2021	
	Total	Total	Total	Total	NTT East ･West (Re-entry)			Total	Total	Total	Total	NTT East ･ West (Re-entry)
					Total	Business	Residential					
Hokkaido	116,055	106,018	96,904	88,688	69,707	61,506	8,201	787	744	648	601	333
Aomori	22,536	20,891	19,441	17,671	13,559	12,824	735	118	112	112	100	70
Iwate	24,172	22,302	20,915	19,342	14,686	13,834	852	97	88	83	71	47
Miyagi	48,994	45,352	41,782	38,650	26,902	25,400	1,502	449	431	420	299	143
Akita	18,294	16,895	15,922	14,615	11,528	10,860	668	97	93	90	82	63
Yamagata	19,657	18,122	16,569	15,009	11,646	10,949	697	93	93	90	76	48
Fukushima	33,518	31,015	28,605	26,197	20,396	18,856	1,540	125	118	105	97	57
Ibaraki	46,688	42,538	39,402	36,132	26,572	24,834	1,738	250	219	214	208	136
Tochigi	34,712	31,698	29,360	26,501	19,190	17,708	1,482	263	252	242	228	162
Gunma	34,139	31,425	29,164	26,509	19,123	17,551	1,572	228	229	221	213	131
Saitama	117,783	108,487	101,027	91,635	57,872	52,137	5,735	898	918	862	841	354
Chiba	100,981	92,803	85,720	77,725	52,411	48,239	4,172	1,028	945	865	759	426
Tokyo	476,007	440,386	400,743	363,433	227,757	210,838	16,919	15,873	15,248	14,562	13,488	5,595
Kanagawa	167,789	156,573	144,260	131,599	86,646	78,960	7,686	2,668	2,549	2,436	2,036	989
Niigata	41,720	38,380	35,639	32,556	23,813	22,500	1,313	160	148	141	132	74
Toyama	22,293	20,319	18,538	17,111	13,328	12,275	1,053	159	149	135	129	72
Ishikawa	24,520	22,508	20,698	19,243	14,921	13,657	1,264	186	180	175	160	77
Fukui	15,667	14,269	13,162	12,107	9,865	9,265	600	75	71	66	62	52
Yamanashi	15,409	14,316	13,181	11,999	9,632	8,821	811	78	75	70	61	46
Nagano	41,981	38,466	35,286	31,594	24,323	21,942	2,381	200	189	170	148	65
Gifu	39,703	36,506	33,747	31,176	24,625	22,419	2,206	197	166	162	138	82
Shizuoka	73,513	67,137	61,060	55,986	39,490	37,381	2,109	386	377	343	321	197
Aichi	152,646	140,621	129,553	118,963	81,893	76,361	5,532	1,371	1,342	1,226	1,120	627
Mie	36,363	33,548	31,212	28,681	23,370	21,532	1,838	170	155	143	115	76
Shiga	25,892	23,739	21,818	20,267	15,573	14,450	1,123	144	137	121	108	49
Kyoto	54,208	49,791	45,485	42,334	29,095	26,051	3,044	341	336	324	294	162
Osaka	232,199	214,062	197,113	182,486	110,883	102,308	8,575	3,972	3,847	3,765	3,412	1,596
Hyogo	88,503	82,250	76,196	71,007	51,462	47,625	3,837	779	760	743	685	356
Nara	20,836	19,194	17,713	16,594	11,870	10,212	1,658	95	90	89	81	53
Wakayama	16,323	15,010	13,803	12,801	10,271	9,387	884	63	69	64	65	47
Tottori	12,032	11,182	10,344	9,682	8,385	7,686	699	54	52	46	41	30
Shimane	15,405	14,423	13,431	12,438	11,016	10,112	904	128	122	117	103	49
Okayama	40,761	37,761	35,287	33,179	26,193	24,066	2,127	221	210	191	178	128
Hiroshima	63,269	58,886	54,292	50,887	38,976	35,907	3,069	365	340	320	303	182
Yamaguchi	28,633	26,520	24,268	22,780	18,645	16,990	1,655	128	131	103	98	64
Tokushima	14,429	13,383	12,251	11,452	9,492	8,737	755	71	59	57	51	35
Kagawa	21,397	19,519	18,086	16,691	12,665	12,004	661	148	143	130	124	70
Ehime	25,832	23,655	21,325	19,702	16,315	15,042	1,273	152	142	123	123	69
Kochi	14,955	13,962	12,947	12,171	10,536	9,859	677	74	73	68	64	53
Fukuoka	111,003	102,674	94,743	87,619	56,913	53,534	3,379	1,068	1,008	921	692	325
Saga	13,904	12,970	11,951	11,045	8,715	8,087	628	60	56	54	45	38
Nagasaki	25,234	23,388	21,362	19,561	15,645	14,605	1,040	152	151	141	142	73
Kumamoto	32,442	30,041	27,381	25,302	19,948	18,733	1,215	183	162	143	131	69
Oita	25,078	23,144	21,683	20,426	16,286	15,100	1,186	97	89	81	79	42
Miyazaki	20,156	18,461	16,878	15,623	12,470	11,624	846	118	105	108	97	58
Kagoshima	30,663	28,422	25,802	23,369	18,953	17,824	1,129	123	121	118	114	70
Okinawa	22,202	20,665	19,043	17,691	13,703	13,264	439	252	232	223	204	106
Nationwide	**2,680,496**	**2,473,677**	**2,275,092**	**2,088,229**	**1,467,265**	**1,353,856**	**113,409**	**34,744**	**33,326**	**31,631**	**28,719**	**13,646**

*Compiled by TCA based on data publicized by the Ministry of Internal Affairs and Communications and other organizations

2-1-4 Trends in Number of Mobile Phone and PHS Contracts by Prefecture

(Contracts)

Pref.	FY2018	FY2019	FY2020	FY2021
Hokkaido	5,895,707	5,819,753	5,975,105	6,011,788
Aomori	1,193,077	1,176,981	1,193,270	1,206,927
Iwate	1,168,610	1,150,198	1,171,489	1,186,412
Miyagi	2,680,955	2,795,336	2,957,708	2,561,810
Akita	918,106	899,429	908,889	913,914
Yamagata	1,039,742	1,024,110	1,041,223	1,052,717
Fukushima	1,868,427	1,838,020	1,859,929	1,869,178
Ibaraki	2,912,004	2,856,172	2,899,444	2,942,238
Tochigi	1,959,606	1,944,132	1,985,280	2,001,112
Gunma	2,020,847	1,981,904	2,028,492	2,059,479
Saitama	7,896,874	7,686,590	7,901,584	8,060,656
Chiba	6,654,827	6,544,681	6,761,478	6,902,491
Tokyo	53,622,797	60,034,916	62,247,537	66,686,306
Kanagawa	10,362,330	10,149,863	10,864,406	11,288,054
Niigata	2,171,151	2,133,268	2,164,965	2,187,020
Toyama	1,089,369	1,082,649	1,131,203	1,175,890
Ishikawa	1,190,816	1,179,718	1,208,789	1,271,549
Fukui	785,987	770,213	787,995	797,662
Yamanashi	852,212	830,699	841,432	853,072
Nagano	2,209,218	2,509,160	3,284,352	4,376,423
Gifu	2,029,266	1,990,436	2,092,344	2,139,534
Shizuoka	3,859,571	3,814,373	3,946,736	4,077,015
Aichi	9,617,688	9,871,726	10,383,697	10,671,784
Mie	1,821,398	1,781,566	1,832,072	1,860,003
Shiga	1,388,804	1,365,235	1,406,632	1,436,902
Kyoto	2,848,874	2,801,816	2,891,224	2,962,949
Osaka	11,562,119	11,585,950	12,229,891	12,617,342
Hyogo	5,672,086	5,531,958	5,726,188	5,811,531
Nara	1,341,371	1,321,433	1,367,343	1,403,451
Wakayama	943,434	920,099	929,237	933,927
Tottori	547,967	533,619	541,380	545,606
Shimane	670,166	657,315	668,920	675,109
Okayama	1,976,981	1,929,221	1,970,231	1,999,420
Hiroshima	3,355,221	3,373,136	3,550,125	3,706,437
Yamaguchi	1,399,108	1,383,085	1,416,291	1,433,223
Tokushima	730,036	717,519	730,836	745,077
Kagawa	1,046,049	1,020,433	1,034,491	1,042,825
Ehime	1,394,763	1,376,297	1,414,327	1,436,236
Kochi	699,776	685,580	695,020	697,685
Fukuoka	9,278,106	10,316,489	11,669,800	12,299,166
Saga	804,274	787,075	809,684	819,223
Nagasaki	1,331,605	1,301,392	1,333,284	1,342,954
Kumamoto	1,787,918	1,755,511	1,837,404	1,861,362
Oita	1,147,839	1,135,313	1,151,247	1,159,468
Miyazaki	1,057,817	1,042,396	1,062,780	1,073,284
Kagoshima	1,577,438	1,545,044	1,568,619	1,587,702
Okinawa	1,490,457	1,562,300	1,580,520	1,591,049
Total	**179,872,794**	**186,514,109**	**195,054,893**	**203,334,962**

*Compiled by TCA based on data publicized by the Ministry of Internal Affairs and Communications

2-1-5 Trends in Number of Domestic Leased Circuits

(Thousand circuits)

	FY2017	FY2018	FY2019	FY2020	FY2021
General Leased Circuits (Frequency Band Use)	203	197	192	191	183
General Leased Circuits (Code Transmission)	20	19	18	17	17
High-Speed Digital Transmission Services	109	78	43	42	37

*Compiled by TCA based on data publicized by the Ministry of Internal Affairs and Communications

2-1-6 Trends in Number of Broadband Service Contracts, etc.

(Contracts)

		FY2019	FY2020	FY2021	FY2022
Internet connection service (for fixed communication)	(total of 54 providers)	41,919,164	42,721,659	43,155,633	41,943,435
Internet connection service (for mobile communication)	(total of 29 providers)	185,242,351	191,334,287	196,516,577	197,637,976
FTTH access service	(total of 310 providers)	33,175,212	35,157,536	36,905,039	38,065,163
DSL access service	(total of 10 providers)	1,397,840	1,073,135	689,816	356,891
CATV access service	(total of 214 providers)	6,675,425	6,534,902	6,404,881	6,277,110
FWA access service	(total of 21 providers)	4,343	3,549	3,111	1,456
BWA access service	(total of 107 providers)	71,200,466	75,708,966	79,731,989	84,276,055
3.9-4G mobile phone terminals packet communications service	(total of 5 providers)	152,623,405	154,366,473	139,054,534	127,379,501
5G mobile phone terminals packet communications service	(total of 5 providers)	24,040	14,185,509	45,018,488	69,808,822
Local 5G service	(total of 12 providers)	–	0	49	136
Mobile Phone and PHS terminal Internet connection service	(total of 5 providers)	186,310,026	194,935,826	203,269,615	210,702,213
Public radio LAN access service	(total of 19 providers)	119,071,867	125,051,323	101,005,848	99,720,918
IP-VPN service	(total of 49 providers)	659,281	660,041	660,218	655,856
Wide-area Ethernet service	(total of 83 providers)	643,824	662,529	678,420	697,439

*Compiled by TCA based on data publicized by the Ministry of Internal Affairs and Communications

2-2 Situation of Traffic

2-2-1 Situation of Total Traffic

2-2-1-1 Trends in Total Number of Calls

(100 Million calls)

Outgoing \ Incoming	Subscriber Telephoe/ISDN					IP Phone				
	FY2017	FY2018	FY2019	FY2020	FY2021	FY2017	FY2018	FY2019	FY2020	FY2021
Subscriber Telephone	76.9	65.8	53.8	42.3	37.3					
Public Telephone	0.7	0.6	0.5	0.4	0.3	1.4	1.3	1.2	1.2	1.2
ISDN	72.9	63.8	57.3	47.3	42.1					
IP Phone	120.2	121.5	121.1	110.2	108.7	11.5	12.1	12.0	11.3	13.5
Mobile Phone/PHS	56.6	50.5	45.6	39.6	37.9	70.5	72.0	72.3	69.9	71.7
Total	**327.3**	**302.2**	**278.2**	**239.7**	**226.3**	**83.4**	**85.4**	**85.5**	**82.4**	**86.4**

Outgoing \ Incoming	Mobile Phone/PHS					Total				
	FY2017	FY2018	FY2019	FY2020	FY2021	FY2017	FY2018	FY2019	FY2020	FY2021
Subscriber Telephone										
Public Telephone	23.0	21.2	19.5	17.4	16.3	174.9	152.7	132.2	108.6	97.2
ISDN										
IP Phone	29.2	30.4	31.3	32.1	34.8	160.9	164.0	164.3	153.5	157.0
Mobile Phone/PHS	358.9	343.8	327.4	307.1	302.8	486.1	466.3	445.3	416.5	412.4
Total	**411.1**	**395.5**	**378.1**	**356.5**	**353.9**	**821.8**	**783.0**	**741.8**	**678.7**	**666.6**

*Compiled by TCA based on data publicized by the Ministry of Internal Affairs and Communications

2-2-1-2 Trends in Total Number of Calls between Fixed Telephone and Mobile Telephone

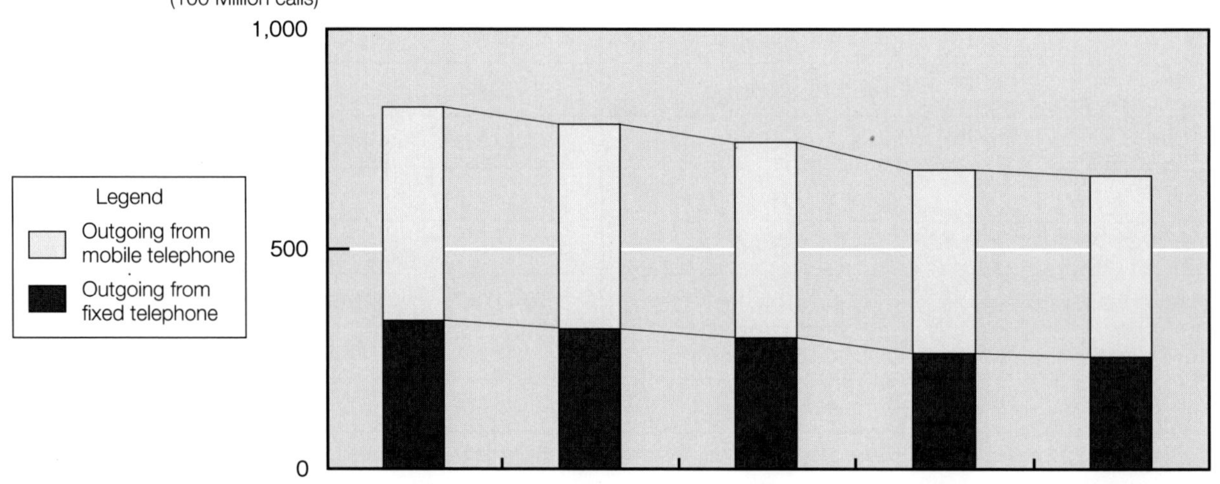

(100 Million calls)

Outgoing	Incoming	FY2017	FY2018	FY2019	FY2020	FY2021
Fixed	Fixed	283.6	265.1	245.9	212.7	203.1
Fixed	Mobile	52.2	51.6	50.8	49.5	51.1
Mobile	Mobile	358.9	343.8	327.4	307.1	302.8
Mobile	Fixed	127.1	122.5	117.9	109.5	109.6
Total		**821.8**	**783.0**	**741.8**	**678.7**	**666.6**

Note: Outgoing from fixed telephone: Outgoing from subscriber telephones, public telephones, ISDN and IP phones
Outgoing from mobile telephone: Outgoing from mobile phones and PHS
Incoming to fixed telephone: Incoming to subscriber telephones, ISDN and IP phones
Incoming to mobile telephone: Incoming to mobile phones and PHS

*Compiled by TCA based on data publicized by the Ministry of Internal Affairs and Communications

2-2-1-3 Trends in Daily Number of Calls per Subscription (Contract)

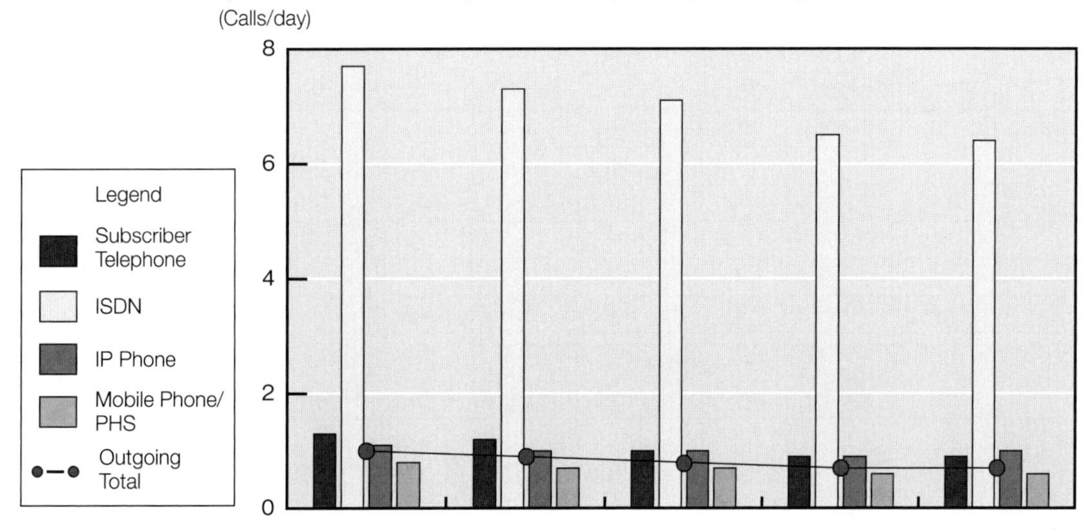

(Calls / day)

Outgoing	FY2017	FY2018	FY2019	FY2020	FY2021
Subscriber Telephone	1.3	1.2	1.0	0.9	0.9
ISDN	7.7	7.3	7.1	6.5	6.4
IP Phone	1.1	1.0	1.0	0.9	1.0
Mobile Phone/PHS	0.8	0.7	0.7	0.6	0.6
Outgoing Total	**1.0**	**0.9**	**0.8**	**0.7**	**0.7**

Note: The categories of respective outgoing calls are as listed below. For example, the number of outgoing calls from subscriber telephones shows the total number of calls outgoing from subscriber telephones and destined for fixed telephones, IP phones, mobile phones, and PHS terminals. Since the actual number of outgoing calls from fixed telephones and destined for IP phones, mobile phones and PHS terminals cannot be identified, the number of those calls is calculated according to the ratio to the number of outgoing calls from fixed telephones and destined for fixed telephones.

Outgoing	ISDN	Cellular Telephone	PHS
Incoming	Fixed Telephone, IP Phone, Mobile Phone, PHS	Fixed Telephone, IP Phone, Mobile Phone, PHS	Fixed Telephone, IP Phone, Mobile Phone, PHS

*Compiled by TCA based on data publicized by the Ministry of Internal Affairs and Communications

2-2-1-4 Trends in Total Call Duration

(Million hours)

Outgoing \ Incoming	Subscriber Telephoe/ISDN					IP Phone				
	FY2017	FY2018	FY2019	FY2020	FY2021	FY2017	FY2018	FY2019	FY2020	FY2021
Subscriber Telephone	234.3	194.6	154.3	130.1	111.6	5.0	4.4	4.2	4.3	4.1
Public Telephone	1.5	1.3	1.1	1.0	0.9					
ISDN	169.6	153.3	138.4	115.2	100.7					
IP Phone	351.7	340.4	327.5	304.2	286.8	48.3	49.9	48.2	48.7	52.4
Mobile Phone/PHS	201.5	194.6	183.9	183.9	178.6	256.3	276.5	303.2	334.1	355.0
Total	**958.6**	**884.1**	**805.2**	**734.3**	**678.6**	**309.6**	**330.8**	**355.6**	**387.1**	**411.5**

Outgoing \ Incoming	Mobile Phone/PHS					Total				
	FY2017	FY2018	FY2019	FY2020	FY2021	FY2017	FY2018	FY2019	FY2020	FY2021
Subscriber Telephone	67.6	63.3	59.3	60.3	55.5	478.0	416.9	357.3	310.9	272.8
Public Telephone										
ISDN										
IP Phone	89.3	93.6	97.8	114.1	121.4	489.2	483.9	473.5	466.9	460.7
Mobile Phone/PHS	1,722.6	1,656.1	1,607.1	1,736.2	1,707.5	2,180.4	2,127.2	2,094.2	2,254.2	2,241.1
Total	**1,879.4**	**1,813.0**	**1,764.2**	**1,910.6**	**1,884.5**	**3,147.6**	**3,027.9**	**2,925.0**	**3,032.1**	**2,974.6**

*Compiled by TCA based on data publicized by the Ministry of Internal Affairs and Communications

2-2-1-5 Trends in Average Call Duration per Call

(Seconds)

Outgoing \ Incoming	Subscriber Telephoe/ISDN					IP Phone				
	FY2017	FY2018	FY2019	FY2020	FY2021	FY2017	FY2018	FY2019	FY2020	FY2021
Subscriber Telephone	109.7	106.5	103.2	110.7	107.7	128.6	121.8	126.0	129.0	123.0
Public Telephone	77.1	78.0	79.2	90.0	108.0					
ISDN	83.8	86.5	87.0	87.7	86.1					
IP Phone	105.3	100.9	97.4	99.4	95.0	151.2	148.5	144.6	155.2	139.7
Mobile Phone/PHS	128.2	138.7	145.2	167.2	169.6	130.9	138.3	151.0	172.1	178.2
Total	**105.4**	**105.3**	**104.2**	**110.3**	**108.0**	**133.6**	**139.4**	**149.7**	**169.1**	**171.5**

Outgoing \ Incoming	Mobile Phone/PHS					Total				
	FY2017	FY2018	FY2019	FY2020	FY2021	FY2017	FY2018	FY2019	FY2020	FY2021
Subscriber Telephone	105.8	107.5	109.5	124.8	122.6	98.4	98.3	97.3	103.1	101.0
Public Telephone										
ISDN										
IP Phone	110.1	110.8	112.5	128.0	125.6	109.5	106.2	103.7	109.5	105.6
Mobile Phone/PHS	172.8	173.4	176.7	203.5	203.0	161.5	164.2	169.3	194.8	195.6
Total	**164.6**	**165.0**	**168.0**	**192.9**	**191.7**	**137.9**	**139.2**	**142.0**	**160.8**	**160.6**

Note: Total Call Duration (seconds) ÷ Total Number of Calls (calls)
*Compiled by TCA based on data publicized by the Ministry of Internal Affairs and Communications

2-2-1-6 Trends in Daily Call Duration per Subscription (Contract)

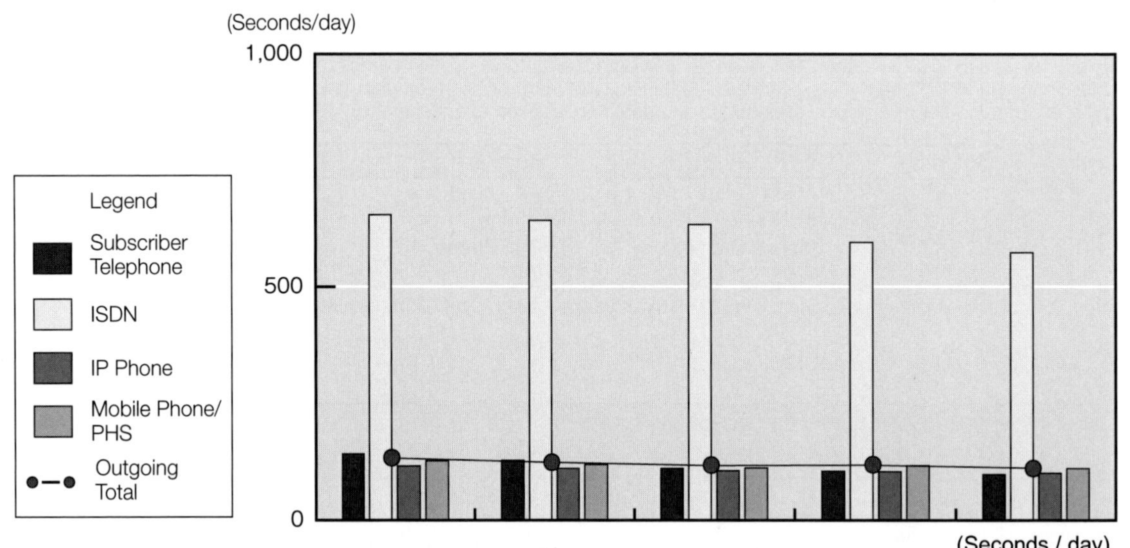

(Seconds / day)

Outgoing	FY2017	FY2018	FY2019	FY2020	FY2021
Subscriber Telephone	142	128	111	105	98
ISDN	655	643	634	596	574
IP Phone	116	111	106	104	101
Mobile Phone/PHS	127	119	112	117	111
Outgoing Total	**133**	**124**	**117**	**118**	**112**

Note: The category of outgoing call duration and calculation method are the same as those in note of 2-2-1-3.
*Compiled by TCA based on data publicized by the Ministry of Internal Affairs and Communications

2-2-2 Situation of Traffic of Subscriber Telephone/ISDN

2-2-2-1 Situation of Calls by Time Zone

2-2-2-1-1 Trends in Number of Calls by Time Zone

(Million calls)

Time Zone	FY2017	FY2018	FY2019	FY2020	FY2021
0-1	116	100	87	71	63
1-2	102	89	79	66	58
2-3	92	81	71	61	55
3-4	84	75	67	58	52
4-5	86	76	68	60	55
5-6	107	93	81	73	65
6-7	148	130	113	97	87
7-8	283	244	202	164	150
8-9	697	616	509	415	371
9-10	1,454	1,267	1,085	869	768
10-11	1,518	1,323	1,132	919	816
11-12	1,406	1,227	1,055	862	763
12-13	852	733	626	519	462
13-14	1,242	1,074	925	757	673
14-15	1,247	1,082	932	767	676
15-16	1,245	1,077	933	768	678
16-17	1,244	1,083	939	760	671
17-18	1,048	905	774	597	525
18-19	714	602	503	381	334
19-20	493	410	344	260	228
20-21	322	267	226	175	154
21-22	201	169	144	109	95
22-23	149	128	109	82	73
23-24	128	111	95	74	66
Total	**14,975**	**12,961**	**11,103**	**8,966**	**7,938**

Number of Calls by Time Zone (FY2021)

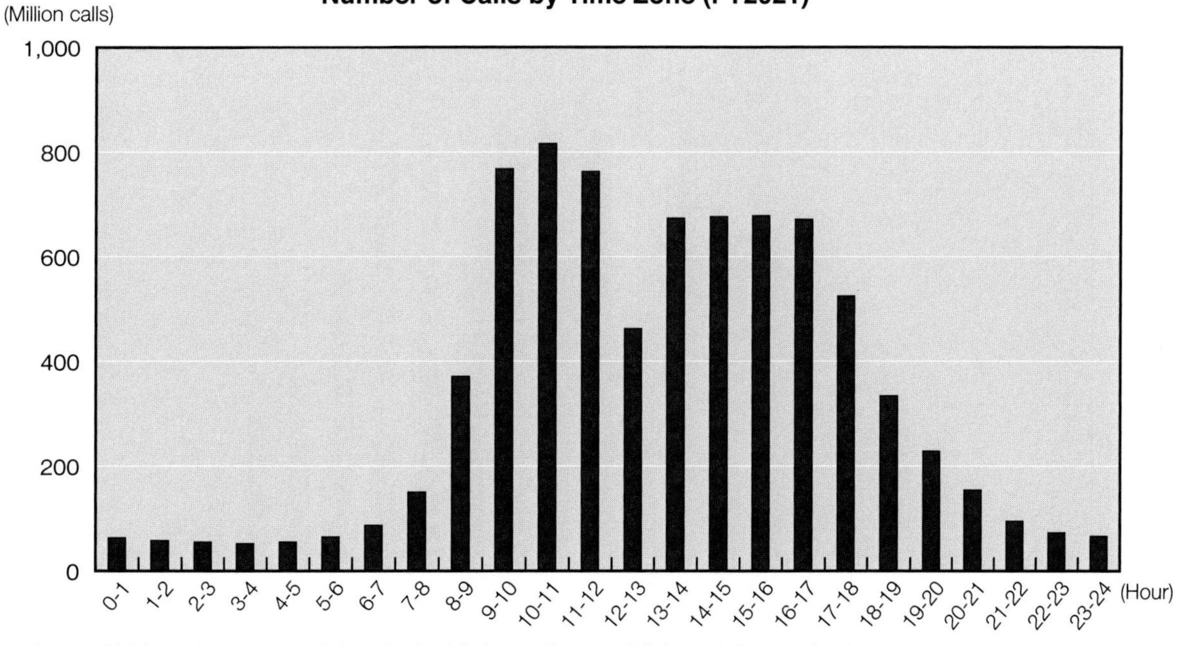

*Compiled by TCA based on data publicized by the Ministry of Internal Affairs and Communications

2-2-2-1-2 Trends in Call Duration by Time Zone

(Million hours)

Time Zone	FY2017	FY2018	FY2019	FY2020	FY2021
0-1	1.59	1.27	1.07	0.78	0.75
1-2	1.20	1.01	0.87	0.67	0.57
2-3	1.00	0.84	0.74	0.57	0.49
3-4	1.26	1.13	1.01	0.79	0.68
4-5	1.03	0.90	1.24	0.68	0.56
5-6	1.20	1.00	0.86	0.71	0.63
6-7	1.91	1.68	1.43	1.12	0.97
7-8	5.06	4.26	3.44	2.66	2.36
8-9	15.94	13.87	11.37	9.29	8.10
9-10	39.72	34.44	29.05	24.05	21.06
10-11	41.54	36.28	30.71	26.48	23.03
11-12	36.85	32.32	27.51	23.81	20.73
12-13	22.58	19.66	16.53	14.53	12.67
13-14	34.16	29.78	25.39	22.25	19.36
14-15	34.78	30.35	26.02	22.97	19.88
15-16	35.66	31.15	26.68	23.35	20.28
16-17	36.74	32.02	27.38	23.19	20.19
17-18	29.75	25.50	21.24	16.79	14.42
18-19	21.15	17.60	14.09	10.81	9.17
19-20	17.15	13.99	11.08	8.64	7.25
20-21	12.42	9.98	7.85	6.05	4.92
21-22	6.21	4.96	3.90	2.88	2.31
22-23	3.00	2.40	1.94	1.31	1.09
23-24	1.93	1.56	1.27	0.90	0.77
Total	**403.85**	**347.90**	**292.71**	**245.27**	**212.29**

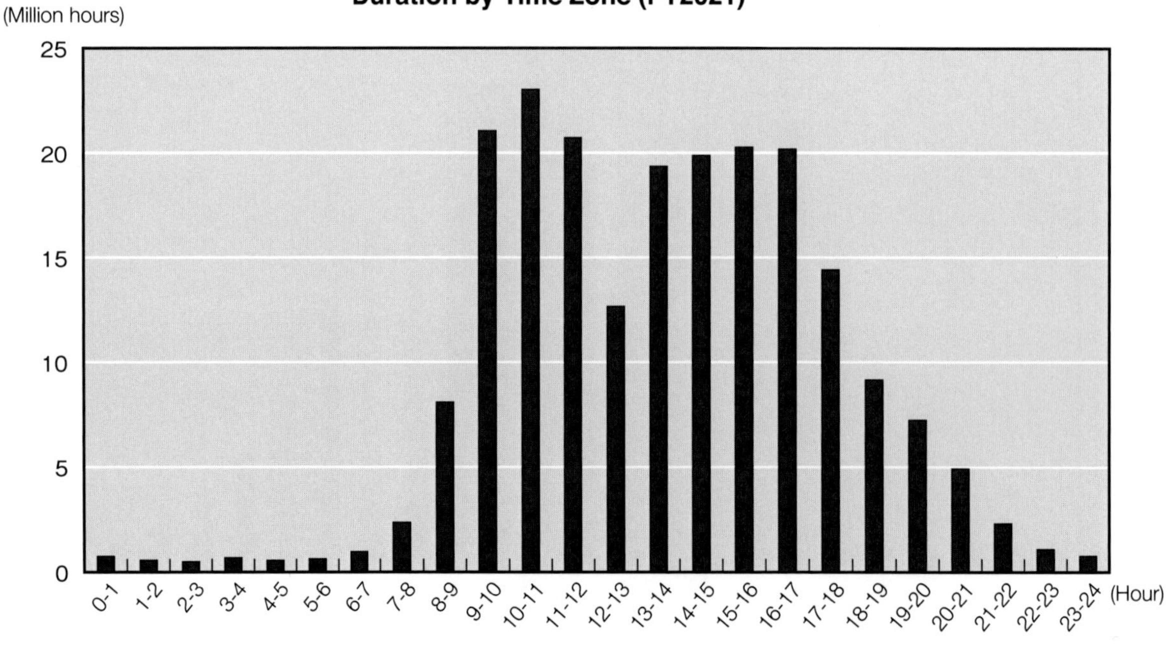

*Compiled by TCA based on data publicized by the Ministry of Internal Affairs and Communications

2-2-2-2 Situation of Number of Calls by Duration

2-2-2-2-1 Trends in Number of Calls by Duration

(Million calls)

Duration	FY2017	FY2018	FY2019	FY2020	FY2021
up to 1 min	10,064	8,709	7,515	6,122	5,442
1-3 mins	3,217	2,798	2,364	1,828	1,601
over 3 mins	1,693	1,454	1,225	1,019	892
Total	**14,975**	**12,961**	**11,103**	**8,966**	**7,938**

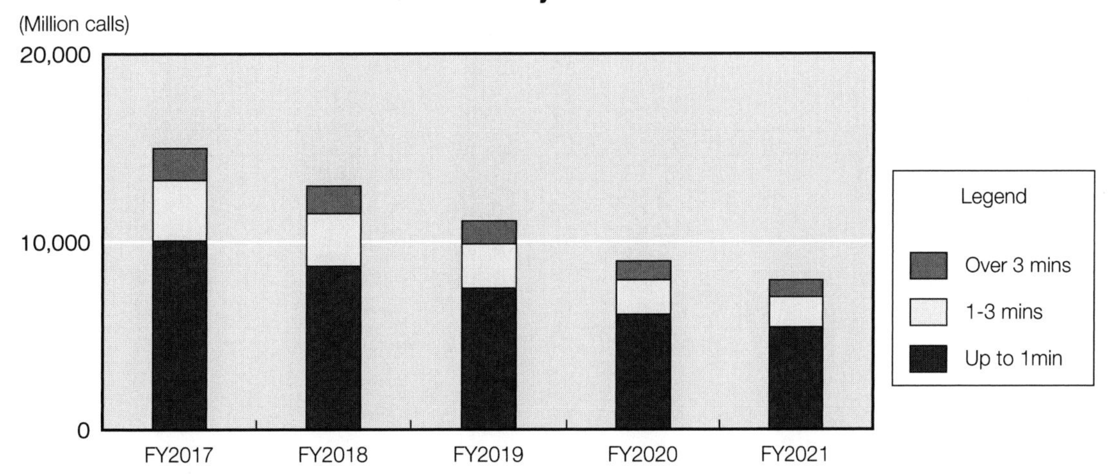

*Compiled by TCA based on data publicized by the Ministry of Internal Affairs and Communications

2-2-2-2-2 Number of Calls by Duration (10-second steps) (FY2021)

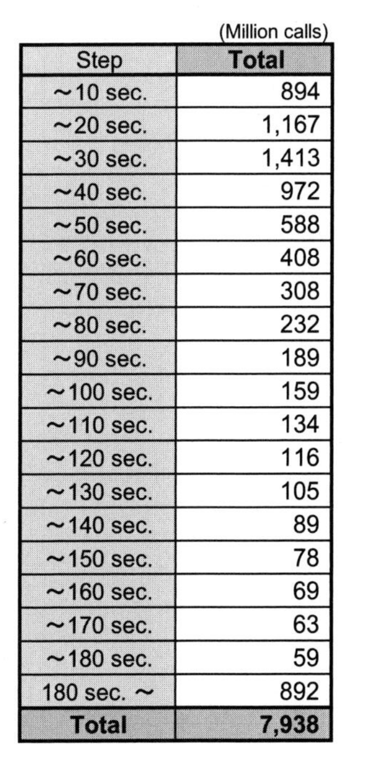

(Million calls)

Step	Total
~10 sec.	894
~20 sec.	1,167
~30 sec.	1,413
~40 sec.	972
~50 sec.	588
~60 sec.	408
~70 sec.	308
~80 sec.	232
~90 sec.	189
~100 sec.	159
~110 sec.	134
~120 sec.	116
~130 sec.	105
~140 sec.	89
~150 sec.	78
~160 sec.	69
~170 sec.	63
~180 sec.	59
180 sec. ~	892
Total	**7,938**

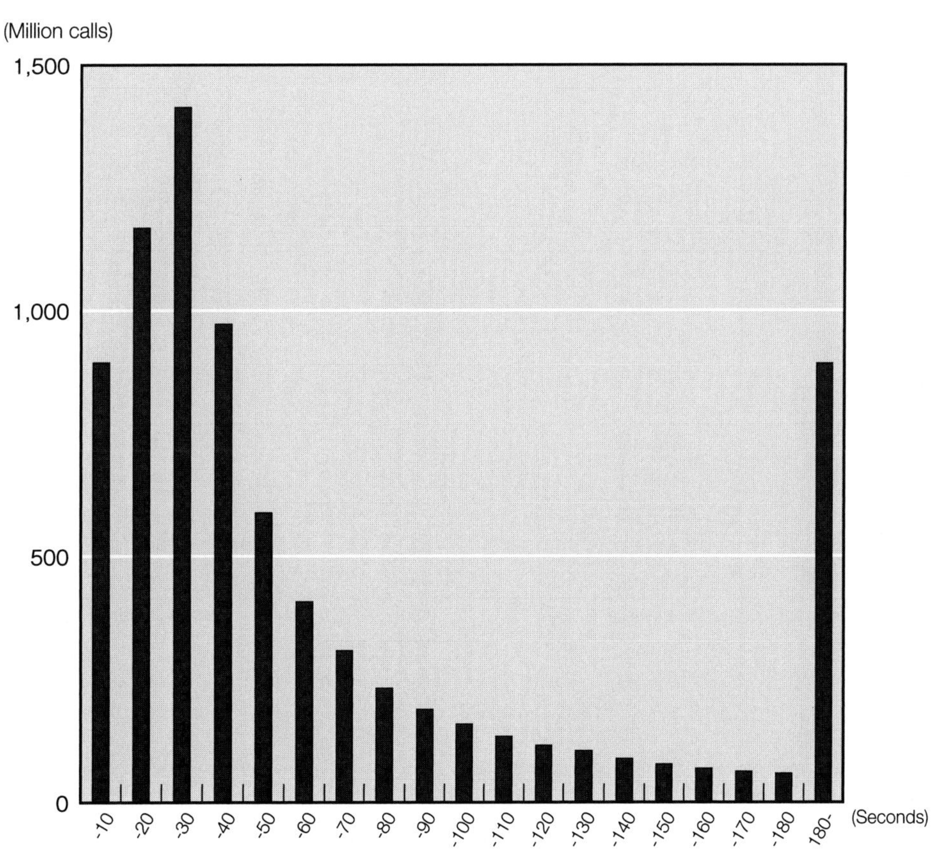

*Compiled by TCA based on data publicized by the Ministry of Internal Affairs and Communications

2-2-2-3 Situation of Calls by Prefecture

2-2-2-3-1 Ranking of Number of Outgoing and Incoming Calls by Prefecture (FY2021)

(Million calls)

Ranking	Outgoing			Incoming		
	Pref.	Number of outgoing calls	Ratio (%)	Pref.	Number of incoming calls	Ratio (%)
1	Tokyo	1,521	19.2	Tokyo	1,455	18.3
2	Osaka	777	9.8	Osaka	774	9.7
3	Kanagawa	528	6.6	Kanagawa	469	5.9
4	Aichi	442	5.6	Aichi	446	5.6
5	Saitama	404	5.1	Saitama	354	4.5
6	Hokkaido	328	4.1	Fukuoka	336	4.2
7	Fukuoka	328	4.1	Chiba	328	4.1
8	Chiba	301	3.8	Hokkaido	327	4.1
9	Hyogo	295	3.7	Hyogo	268	3.4
10	Shizuoka	199	2.5	Shizuoka	204	2.6
11	Hiroshima	170	2.1	Hiroshima	189	2.4
12	Miyagi	151	1.9	Kyoto	174	2.2
13	Kyoto	151	1.9	Miyagi	169	2.1
14	Niigata	130	1.6	Niigata	145	1.8
15	Ibaraki	126	1.6	Ibaraki	126	1.6
16	Nagano	111	1.4	Nagano	126	1.6
17	Gifu	103	1.3	Gifu	111	1.4
18	Okayama	103	1.3	Gunma	107	1.3
19	Fukushima	98	1.2	Okayama	106	1.3
20	Gunma	95	1.2	Fukushima	100	1.3
21	Kagoshima	92	1.2	Tochigi	93	1.2
22	Mie	85	1.1	Mie	89	1.1
23	Kumamoto	84	1.1	Kumamoto	89	1.1
24	Tochigi	84	1.1	Kagoshima	88	1.1
25	Yamaguchi	74	0.9	Iwate	73	0.9
26	Iwate	73	0.9	Yamaguchi	71	0.9
27	Aomori	71	0.9	Aomori	69	0.9
28	Nagasaki	70	0.9	Nagasaki	68	0.9
29	Ishikawa	63	0.8	Ehime	68	0.9
30	Ehime	63	0.8	Ishikawa	67	0.8
31	Shiga	63	0.8	Yamagata	64	0.8
32	Oita	62	0.8	Shiga	62	0.8
33	Yamagata	59	0.7	Oita	61	0.8
34	Kagawa	59	0.7	Akita	60	0.8
35	Akita	57	0.7	Toyama	60	0.8
36	Miyazaki	54	0.7	Okinawa	59	0.7
37	Toyama	54	0.7	Kagawa	59	0.7
38	Okinawa	53	0.7	Miyazaki	56	0.7
39	Nara	53	0.7	Shimane	54	0.7
40	Shimane	49	0.6	Nara	52	0.7
41	Wakayama	48	0.6	Wakayama	47	0.6
42	Kochi	38	0.5	Fukui	40	0.5
43	Yamanashi	36	0.5	Saga	38	0.5
44	Fukui	35	0.4	Kochi	38	0.5
45	Saga	35	0.4	Yamanashi	37	0.5
46	Tokushima	32	0.4	Tokushima	34	0.4
47	Tottori	29	0.4	Tottori	30	0.4
	Total	**7,938**	**100.0**	**Total**	**7,938**	**100.0**

*Compiled by TCA based on data publicized by the Ministry of Internal Affairs and Communications

2-2-2-3-2 Main Destination Prefectures by Originating Prefecture (FY2021)

Outgoing	Total Number of Outgoing calls (million)	Incoming									
		1		2		3		4		5	
		Pref.	Ratio (%)	Pref.	Ratio (%)	Pref.	Ratio (%)	Pref.	Ratio (%)	Pref.	Ratio (%)
Hokkaido	328	Hokkaido	79.0	Tokyo	8.0	Miyagi	2.4	Kanagawa	1.2	Osaka	1.1
Aomori	71	Aomori	74.3	Miyagi	7.6	Tokyo	5.5	Iwate	2.7	Akita	1.3
Iwate	73	Iwate	71.8	Miyagi	9.9	Tokyo	5.7	Aomori	2.3	Akita	1.6
Miyagi	151	Miyagi	62.9	Tokyo	9.2	Fukushima	3.8	Iwate	2.9	Yamagata	2.7
Akita	57	Akita	75.2	Miyagi	6.6	Tokyo	5.6	Yamagata	1.8	Aomori	1.6
Yamagata	59	Yamagata	71.8	Miyagi	9.0	Tokyo	6.6	Kanagawa	1.2	Saitama	1.1
Fukushima	98	Fukushima	69.4	Miyagi	10.0	Tokyo	9.4	Saitama	1.3	Kanagawa	1.2
Ibaraki	126	Ibaraki	56.2	Tokyo	12.4	Chiba	8.0	Saitama	7.4	Nagano	2.5
Tochigi	84	Tochigi	61.3	Tokyo	13.0	Saitama	6.8	Ibaraki	3.3	Gunma	2.9
Gunma	95	Gunma	57.4	Tokyo	12.9	Saitama	6.5	Niigata	4.4	Tochigi	3.3
Saitama	404	Saitama	44.5	Tokyo	19.7	Chiba	5.6	Kanagawa	3.9	Gunma	2.4
Chiba	301	Chiba	59.7	Tokyo	20.4	Saitama	4.1	Kanagawa	2.8	Ibaraki	2.0
Tokyo	1,521	Tokyo	54.9	Kanagawa	6.2	Saitama	5.2	Osaka	4.4	Chiba	3.8
Kanagawa	528	Kanagawa	53.1	Tokyo	21.0	Osaka	2.8	Chiba	2.7	Saitama	2.7
Niigata	130	Niigata	77.0	Tokyo	7.7	Saitama	1.6	Osaka	1.4	Kanagawa	1.2
Toyama	54	Toyama	67.8	Tokyo	5.8	Ishikawa	5.2	Osaka	4.8	Kyoto	3.2
Ishikawa	63	Ishikawa	60.9	Tokyo	9.3	Toyama	5.0	Osaka	4.5	Kyoto	4.0
Fukui	35	Fukui	71.1	Tokyo	5.2	Osaka	5.1	Kyoto	4.0	Ishikawa	3.8
Yamanashi	36	Yamanashi	62.1	Tokyo	13.5	Saitama	6.8	Shizuoka	4.8	Kanagawa	2.9
Nagano	111	Nagano	68.6	Tokyo	9.1	Chiba	4.5	Niigata	4.0	Aichi	2.2
Gifu	103	Gifu	61.6	Aichi	17.2	Tokyo	5.5	Osaka	3.8	Kanagawa	1.1
Shizuoka	199	Shizuoka	70.5	Tokyo	7.8	Aichi	7.5	Osaka	3.0	Kanagawa	2.8
Aichi	442	Aichi	64.9	Tokyo	6.9	Osaka	4.9	Hyogo	3.5	Gifu	3.2
Mie	85	Mie	65.8	Aichi	12.5	Tokyo	5.6	Osaka	5.3	Kanagawa	1.2
Shiga	63	Shiga	52.9	Osaka	16.4	Kyoto	11.3	Tokyo	5.3	Aichi	2.2
Kyoto	151	Kyoto	61.1	Osaka	15.6	Tokyo	5.9	Shiga	2.8	Hyogo	2.4
Osaka	777	Osaka	57.1	Tokyo	7.8	Hyogo	5.4	Kyoto	3.1	Aichi	3.0
Hyogo	295	Hyogo	49.7	Osaka	20.0	Tokyo	6.2	Fukuoka	2.6	Kyoto	1.8
Nara	53	Nara	53.2	Osaka	20.7	Kyoto	9.3	Tokyo	5.4	Hyogo	1.8
Wakayama	48	Wakayama	62.8	Osaka	12.9	Tokyo	6.1	Kyoto	4.4	Aichi	1.6
Tottori	29	Tottori	68.9	Shimane	9.4	Hiroshima	4.5	Osaka	4.1	Tokyo	3.9
Shimane	49	Shimane	61.4	Tokyo	12.2	Hiroshima	5.8	Osaka	5.5	Tottori	2.8
Okayama	103	Okayama	63.3	Hiroshima	8.8	Osaka	6.4	Tokyo	5.2	Hyogo	3.8
Hiroshima	170	Hiroshima	70.2	Osaka	5.2	Tokyo	4.8	Okayama	3.5	Yamaguchi	2.6
Yamaguchi	74	Yamaguchi	65.6	Fukuoka	9.7	Hiroshima	8.1	Tokyo	4.5	Osaka	4.0
Tokushima	32	Tokushima	69.8	Osaka	5.8	Tokyo	5.4	Kagawa	5.2	Hiroshima	3.3
Kagawa	59	Kagawa	61.7	Osaka	6.4	Tokyo	5.5	Ehime	4.6	Hiroshima	3.6
Ehime	63	Ehime	70.2	Tokyo	5.4	Osaka	5.3	Hiroshima	3.9	Kagawa	3.8
Kochi	38	Kochi	75.1	Tokyo	4.7	Osaka	4.3	Kagawa	3.3	Hiroshima	2.6
Fukuoka	328	Fukuoka	64.1	Tokyo	6.1	Osaka	5.1	Kumamoto	2.3	Saga	1.9
Saga	35	Saga	66.8	Fukuoka	16.1	Tokyo	4.2	Nagasaki	2.6	Osaka	2.6
Nagasaki	70	Nagasaki	71.3	Fukuoka	9.9	Tokyo	4.7	Osaka	2.9	Saga	1.6
Kumamoto	84	Kumamoto	70.0	Fukuoka	11.2	Tokyo	4.3	Osaka	2.9	Kagoshima	1.5
Oita	62	Oita	70.8	Fukuoka	11.1	Tokyo	4.1	Osaka	3.0	Hyogo	2.5
Miyazaki	54	Miyazaki	73.4	Fukuoka	7.0	Tokyo	4.1	Kagoshima	2.9	Osaka	2.7
Kagoshima	92	Kagoshima	71.4	Fukuoka	6.1	Tokyo	4.1	Osaka	3.1	Miyazaki	2.3
Okinawa	53	Okinawa	73.9	Tokyo	7.2	Osaka	5.1	Fukuoka	4.6	Kanagawa	1.2

*Compiled by TCA based on data publicized by the Ministry of Internal Affairs and Communications

2-2-2-3-3 Main Originating Prefectures by Destination Prefecture (FY2021)

Incoming	Total number of incoming calls (million)	Outgoing									
		1		2		3		4		5	
		Pref.	**Ratio (%)**	**Pref.**	**Ratio (%)**	**Pref.**	**Ratio (%)**	**Pref.**	**Ratio (%)**	**Pref.**	**Ratio (%)**
Hokkaido	327	Hokkaido	79.3	Tokyo	9.1	Osaka	1.7	Saitama	1.5	Kanagawa	1.3
Aomori	69	Aomori	76.2	Tokyo	6.7	Miyagi	3.8	Iwate	2.5	Saitama	2.3
Iwate	73	Iwate	71.7	Tokyo	7.0	Miyagi	6.0	Aomori	2.7	Saitama	2.4
Miyagi	169	Miyagi	56.2	Tokyo	9.4	Fukushima	5.8	Hokkaido	4.7	Iwate	4.3
Akita	60	Akita	71.2	Tokyo	7.2	Miyagi	3.8	Osaka	3.4	Saitama	2.3
Yamagata	64	Yamagata	66.9	Tokyo	8.0	Miyagi	6.4	Saitama	2.9	Osaka	2.4
Fukushima	100	Fukushima	68.4	Tokyo	9.4	Miyagi	5.7	Saitama	3.2	Chiba	1.7
Ibaraki	126	Ibaraki	55.9	Tokyo	15.2	Saitama	7.3	Chiba	4.9	Kanagawa	3.1
Tochigi	93	Tochigi	54.9	Tokyo	15.0	Saitama	7.4	Kanagawa	3.5	Gunma	3.4
Gunma	107	Gunma	51.2	Tokyo	13.3	Saitama	8.9	Kanagawa	5.8	Osaka	2.8
Saitama	354	Saitama	50.8	Tokyo	22.4	Kanagawa	4.0	Chiba	3.5	Osaka	2.9
Chiba	328	Chiba	54.8	Tokyo	17.7	Saitama	6.9	Kanagawa	4.4	Ibaraki	3.1
Tokyo	1,455	Tokyo	57.4	Kanagawa	7.6	Saitama	5.5	Chiba	4.2	Osaka	4.2
Kanagawa	469	Kanagawa	59.7	Tokyo	19.9	Saitama	3.4	Osaka	2.9	Chiba	1.8
Niigata	145	Niigata	69.0	Tokyo	8.9	Nagano	3.0	Saitama	3.0	Gunma	2.9
Toyama	60	Toyama	61.4	Tokyo	9.0	Osaka	6.5	Ishikawa	5.3	Kanagawa	2.5
Ishikawa	67	Ishikawa	57.8	Tokyo	7.6	Osaka	6.5	Toyama	4.2	Aichi	3.9
Fukui	40	Fukui	62.8	Tokyo	7.9	Osaka	7.1	Ishikawa	4.7	Aichi	2.5
Yamanashi	37	Yamanashi	61.0	Tokyo	16.6	Kanagawa	5.6	Saitama	2.5	Osaka	2.2
Nagano	126	Nagano	60.6	Tokyo	11.1	Osaka	6.2	Aichi	3.5	Ibaraki	2.5
Gifu	111	Gifu	57.5	Aichi	12.7	Tokyo	7.9	Osaka	5.8	Kanagawa	2.3
Shizuoka	204	Shizuoka	68.7	Tokyo	9.9	Aichi	4.8	Kanagawa	3.8	Osaka	3.2
Aichi	446	Aichi	64.3	Tokyo	8.1	Osaka	5.2	Gifu	4.0	Shizuoka	3.4
Mie	89	Mie	62.8	Aichi	11.3	Tokyo	7.8	Osaka	5.7	Kanagawa	1.9
Shiga	62	Shiga	53.9	Osaka	12.6	Tokyo	8.1	Kyoto	6.7	Hyogo	4.2
Kyoto	174	Kyoto	52.7	Osaka	13.8	Tokyo	6.6	Shiga	4.1	Hyogo	3.1
Osaka	774	Osaka	57.4	Tokyo	8.7	Hyogo	7.6	Kyoto	3.0	Aichi	2.8
Hyogo	268	Hyogo	54.7	Osaka	15.8	Tokyo	7.5	Aichi	5.9	Fukuoka	2.1
Nara	52	Nara	53.5	Osaka	19.3	Tokyo	8.2	Hyogo	4.7	Kyoto	2.5
Wakayama	47	Wakayama	64.2	Osaka	13.7	Tokyo	7.5	Hyogo	3.2	Kanagawa	2.0
Tottori	30	Tottori	64.8	Tokyo	6.6	Shimane	4.6	Osaka	4.6	Hyogo	4.5
Shimane	54	Shimane	56.1	Tokyo	9.1	Osaka	5.7	Tottori	5.0	Hiroshima	4.7
Okayama	106	Okayama	61.6	Tokyo	7.5	Osaka	6.2	Hiroshima	5.6	Hyogo	4.7
Hiroshima	189	Hiroshima	63.3	Tokyo	6.5	Okayama	4.8	Osaka	4.3	Yamaguchi	3.2
Yamaguchi	71	Yamaguchi	68.2	Tokyo	6.2	Hiroshima	6.1	Fukuoka	4.9	Osaka	3.8
Tokushima	34	Tokushima	66.3	Tokyo	6.5	Osaka	6.1	Kagawa	5.1	Hyogo	3.4
Kagawa	59	Kagawa	62.0	Tokyo	6.6	Osaka	6.5	Ehime	4.0	Hyogo	3.0
Ehime	68	Ehime	65.5	Tokyo	8.2	Osaka	6.0	Kagawa	4.1	Hyogo	2.4
Kochi	38	Kochi	73.8	Tokyo	5.8	Osaka	4.4	Kagawa	3.2	Hyogo	2.0
Fukuoka	336	Fukuoka	62.6	Tokyo	7.6	Osaka	4.2	Kumamoto	2.8	Hyogo	2.3
Saga	38	Saga	60.5	Fukuoka	16.4	Tokyo	5.8	Saitama	3.3	Osaka	3.1
Nagasaki	68	Nagasaki	73.1	Fukuoka	7.9	Tokyo	6.4	Osaka	2.7	Hyogo	1.7
Kumamoto	89	Kumamoto	66.5	Fukuoka	8.5	Tokyo	7.1	Osaka	3.2	Saitama	2.4
Oita	61	Oita	72.5	Fukuoka	9.4	Tokyo	5.7	Osaka	2.4	Saitama	2.0
Miyazaki	56	Miyazaki	71.0	Tokyo	6.8	Fukuoka	5.4	Kagoshima	3.8	Osaka	2.7
Kagoshima	88	Kagoshima	74.2	Tokyo	6.1	Fukuoka	5.0	Osaka	2.7	Hyogo	2.3
Okinawa	59	Okinawa	66.3	Tokyo	12.3	Osaka	4.7	Fukuoka	3.0	Saitama	2.2

*Compiled by TCA based on data publicized by the Ministry of Internal Affairs and Communications

2-2-2-4 Situation of Share by Carrier in Calls between Prefectures

2-2-2-4-1 Trends in Ratio of Number of Calls by Carrier in Calls between Prefectures

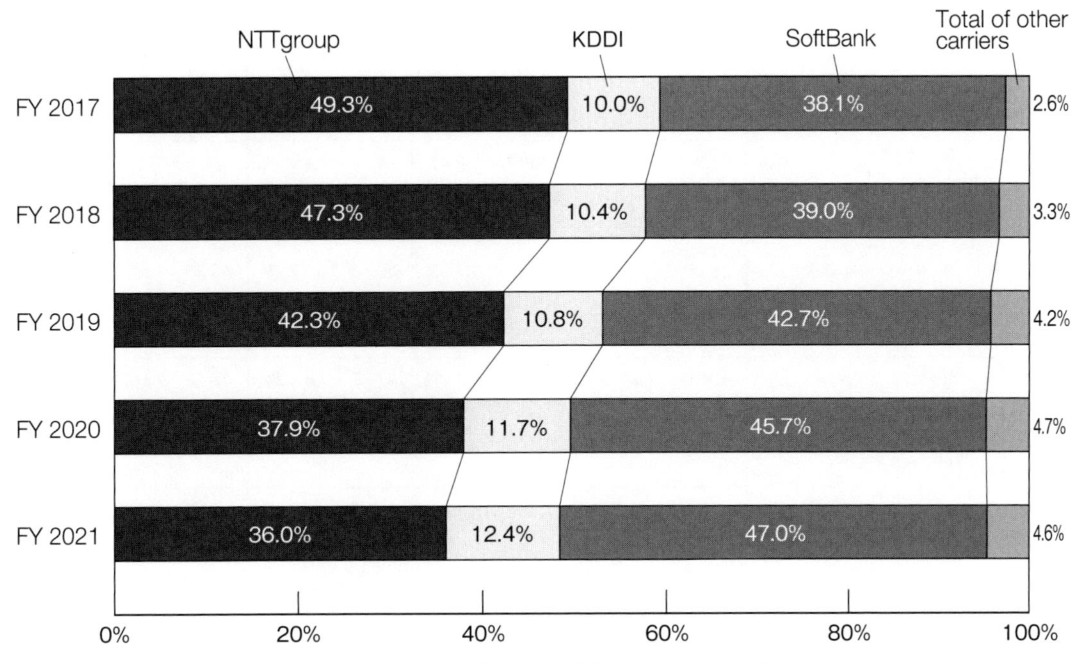

*Compiled by TCA based on data publicized by the Ministry of Internal Affairs and Communications

2-2-2-4-2 Trends in Ratio of Call Hours by Carrier in Calls between Prefectures

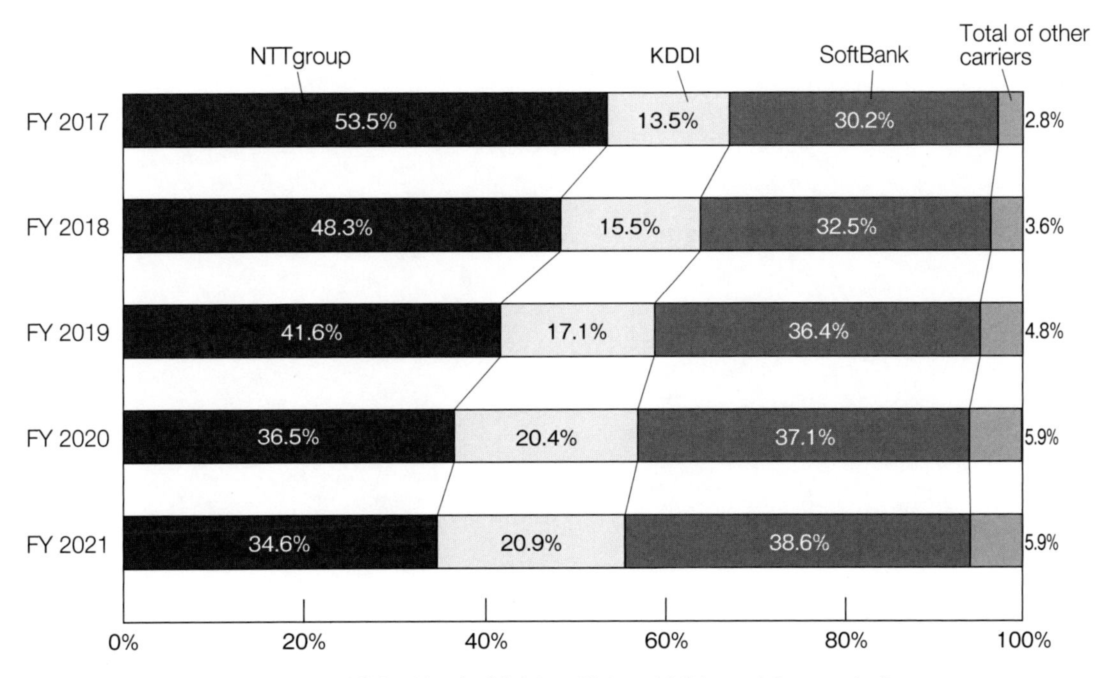

*Compiled by TCA based on data publicized by the Ministry of Internal Affairs and Communications

2-2-3 Situation of Traffic of IP Phones

2-2-3-1 Trends in Number of Telephone Numbers in Use and Communications Traffic

		FY2017		FY2018		FY2019		FY2020		FY2021	
Total number of numbers in use (million numbers)		42.55	(3.8%)	43.41	(2.0%)	44.13	(1.7%)	44.67	(1.2%)	45.35	(1.5%)
	(0ABJ-IP phone)	33.64	(3.7%)	34.46	(2.4%)	35.21	(2.2%)	35.68	(1.3%)	35.94	(0.7%)
	(050-IP phone)	8.91	(4.5%)	8.95	(0.4%)	8.92	(▲0.3%)	8.99	(0.7%)	9.41	(4.7%)
Number of calls (billion calls)		16.23	(3.8%)	16.53	(1.8%)	16.55	(0.1%)	15.47	(▲6.5%)	15.82	(2.3%)
	From IP phones to subscriber telephones, ISDN, IP phones, mobile phones, and PHS phones	16.09	(3.9%)	16.40	(1.9%)	16.43	(0.2%)	15.35	(▲6.6%)	15.70	(2.3%)
	From fixed-line services to IP phones	0.14	(▲11.7%)	0.13	(▲10.9%)	0.12	(▲8.9%)	0.12	(2.4%)	0.12	(5.4%)
Duration of calls (million hours)		494.6	(▲1.0%)	488.5	(▲1.2%)	477.7	(▲2.2%)	471.2	(▲1.4%)	464.7	(▲1.4%)
	From IP phones to subscriber telephones, ISDN, IP phones, mobile phones, and PHS phones	489.5	(▲0.8%)	483.9	(▲1.1%)	473.5	(▲2.1%)	466.9	(▲1.4%)	460.7	(▲1.3%)
	From fixed-line services to IP phones	5.1	(▲12.7%)	4.7	(▲7.9%)	4.2	(▲10.0%)	4.3	(2.3%)	4.1	(▲5.1%)

Notes: Figures in parentheses indicate rates of increase/decrease over the previous fiscal year.
*Compiled by TCA based on data publicized by the Ministry of Internal Affairs and Communications

2-2-4 Situation of Traffic of Mobile and PHS Phones

2-2-4-1 Situation of Calls by Time Zone

2-2-4-1-1 Trends in Number of Calls by Time Zone

(Calls to and from mobile / PHS phones) (Million calls)

Time Zone	FY2017	FY2018	FY2019	FY2020	FY2021
0-1	318	276	248	176	160
1-2	208	181	164	118	108
2-3	152	132	121	90	83
3-4	125	110	100	78	74
4-5	126	112	104	86	82
5-6	204	186	174	149	146
6-7	503	470	440	373	373
7-8	1,188	1,136	1,073	929	943
8-9	2,373	2,317	2,222	2,021	2,053
9-10	3,696	3,638	3,530	3,394	3,431
10-11	3,952	3,877	3,768	3,728	3,711
11-12	3,828	3,739	3,627	3,609	3,569
12-13	3,306	3,170	3,031	2,881	2,838
13-14	3,567	3,474	3,355	3,311	3,271
14-15	3,505	3,420	3,315	3,299	3,260
15-16	3,802	3,706	3,582	3,524	3,497
16-17	4,150	4,036	3,889	3,761	3,742
17-18	4,515	4,328	4,118	3,820	3,755
18-19	3,818	3,586	3,351	2,969	2,890
19-20	2,798	2,586	2,393	2,044	1,995
20-21	2,000	1,824	1,670	1,375	1,334
21-22	1,360	1,224	1,107	857	819
22-23	854	753	679	497	464
23-24	515	447	400	289	263
Total	**50,864**	**48,728**	**46,460**	**43,379**	**42,860**

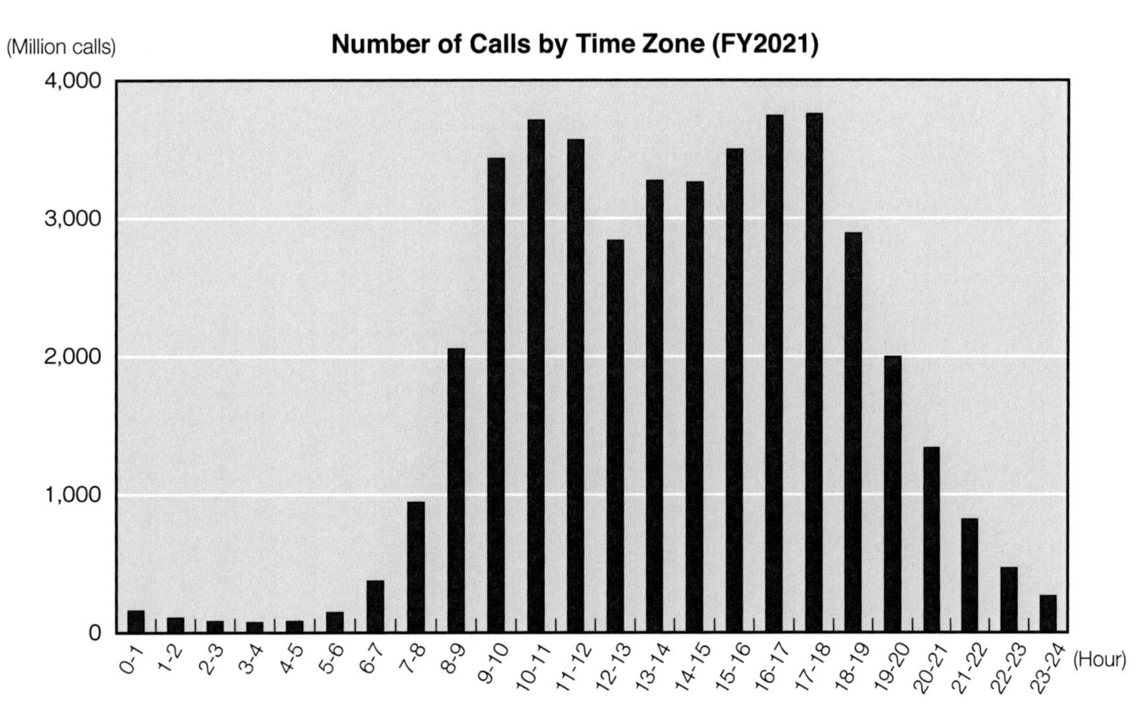

*Compiled by TCA based on data publicized by the Ministry of Internal Affairs and Communications

2-2-4-1-2 Trends in Call Duration by Time Zone

(Calls to and from mobile / PHS phones) (Million hours)

Time Zone	FY2017	FY2018	FY2019	FY2020	FY2021
0-1	37.46	32.19	30.05	28.00	25.83
1-2	23.18	20.03	19.16	18.18	17.24
2-3	15.90	14.11	13.95	13.55	13.11
3-4	12.35	11.28	11.47	11.43	11.36
4-5	12.71	11.47	11.70	11.99	12.05
5-6	13.33	12.79	13.27	13.79	14.18
6-7	22.84	22.32	22.51	22.43	23.01
7-8	46.97	46.37	45.82	44.66	45.83
8-9	87.08	86.83	85.60	85.46	87.42
9-10	138.44	138.73	137.77	147.69	149.96
10-11	149.92	150.16	149.92	168.69	168.10
11-12	140.16	140.20	140.24	160.60	159.23
12-13	122.82	120.95	119.31	128.68	127.89
13-14	129.82	129.57	129.27	147.36	146.35
14-15	131.71	131.97	132.58	154.69	153.52
15-16	143.41	143.68	143.77	165.99	165.41
16-17	157.23	157.15	156.77	177.51	177.74
17-18	173.56	171.32	168.87	182.72	181.49
18-19	160.32	155.87	151.42	158.19	156.67
19-20	141.41	136.20	131.66	136.25	135.73
20-21	131.51	125.06	120.12	124.12	122.23
21-22	111.93	104.41	98.99	99.41	96.21
22-23	84.52	76.19	71.44	68.51	64.94
23-24	58.46	50.97	47.36	44.33	40.94
Total	**2,247.02**	**2,189.83**	**2,153.00**	**2,314.22**	**2,296.47**

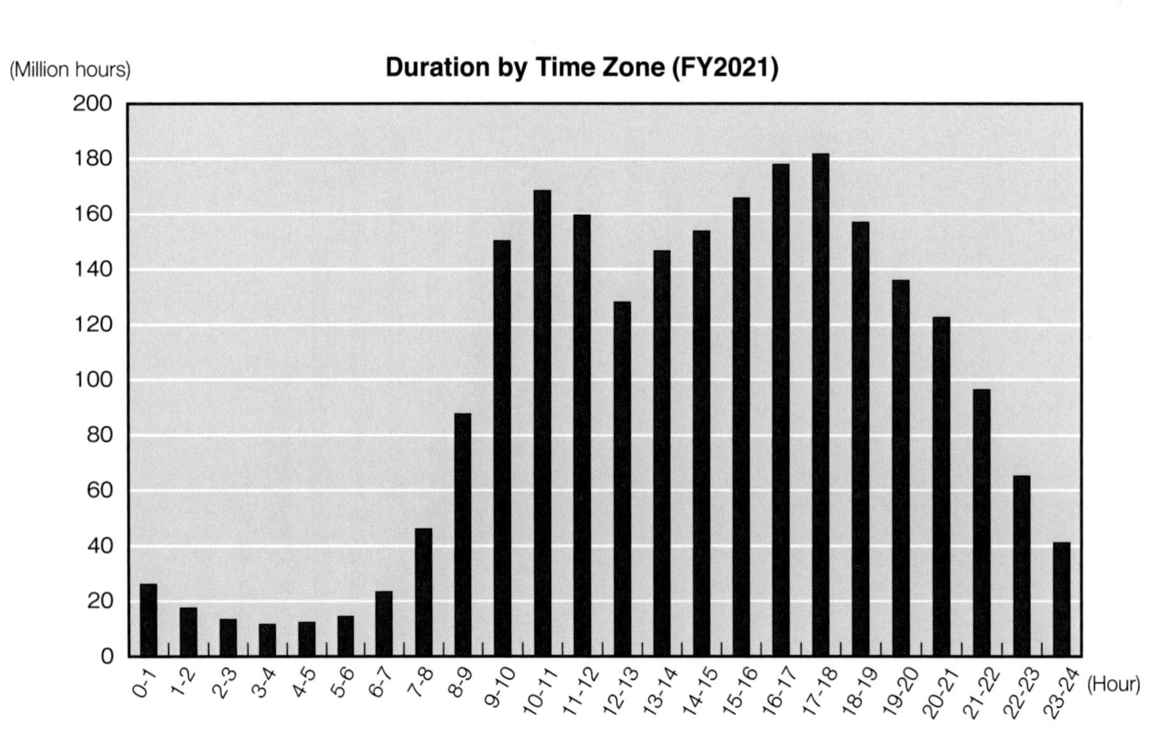

*Compiled by TCA based on data publicized by the Ministry of Internal Affairs and Communications

2-2-4-2 Situation of Number of Calls by Duration

2-2-4-2-1 Trends in Number of Calls by Duration

(Calls to and from mobile / PHS phones) (Million calls)

Duration	FY2017	FY2018	FY2019	FY2020	FY2021
up to 1 min	27,701	26,235	24,894	22,107	21,704
1-3 mins	13,943	13,472	12,804	11,965	11,935
over 3 mins	9,219	9,020	8,763	9,309	9,221
Total	50,864	48,728	46,460	43,379	42,860

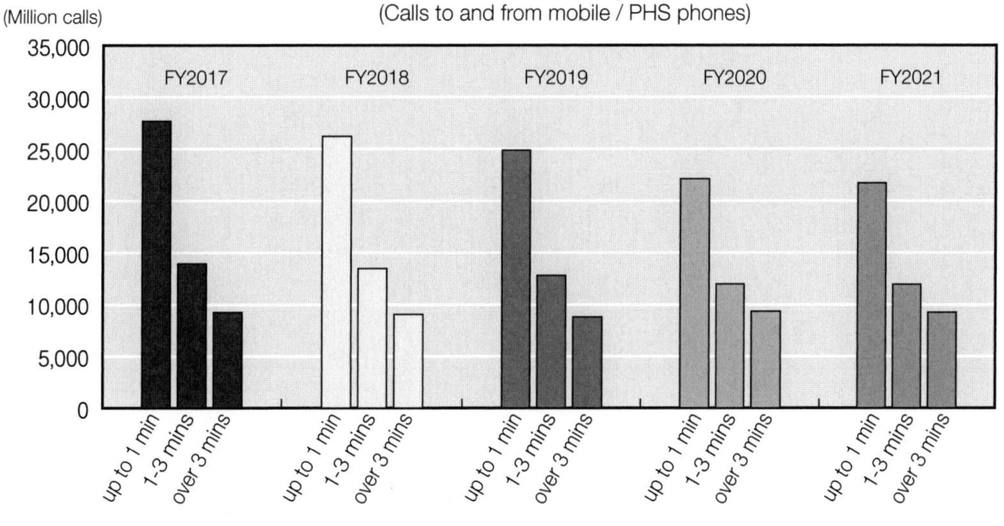

*Compiled by TCA based on data publicized by the Ministry of Internal Affairs and Communications

2-2-4-2-2 Number of Calls by Duration (10-second steps) (FY2021)

(Million calls)

Step	Calls to and from mobile/PHS phones
~10 sec.	4,963
~20 sec.	4,341
~30 sec.	4,072
~40 sec.	3,345
~50 sec.	2,729
~60 sec.	2,254
~70 sec.	1,898
~80 sec.	1,621
~90 sec.	1,392
~100sec.	1,210
~110sec.	1,059
~120sec.	932
~130sec.	825
~140sec.	734
~150sec.	657
~160sec.	590
~170sec.	533
~180sec.	484
180sec.~	9,221
Total	**42,860**

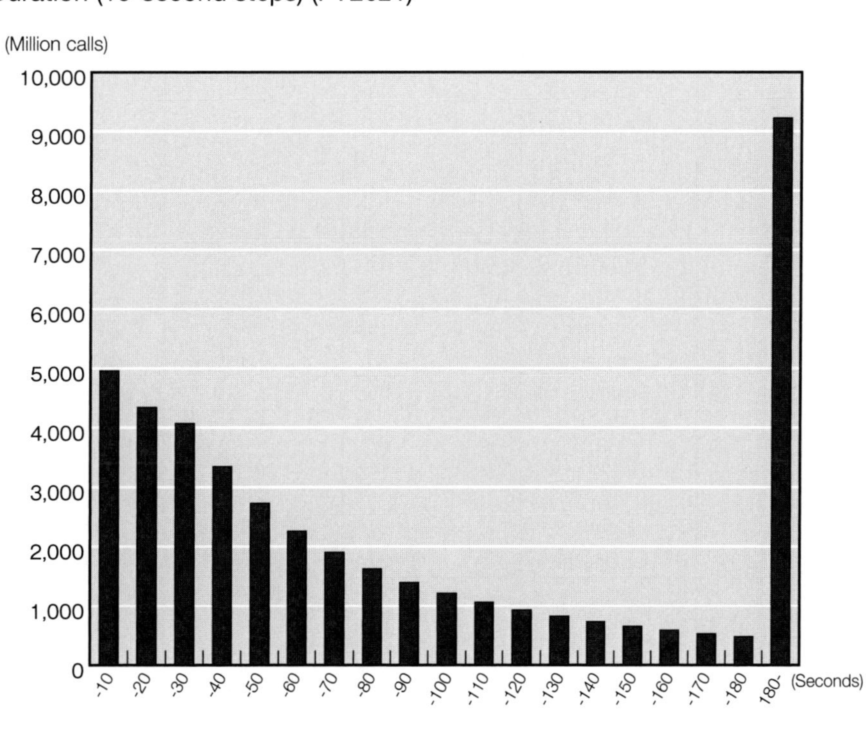

*Compiled by TCA based on data publicized by the Ministry of Internal Affairs and Communications

2-2-4-3 Situation of Calls by Prefecture

2-2-4-3-1 Ranking of Number of Outgoing and Incoming Calls by Prefecture (FY2021)

(Million calls)

Ranking	Outgoing			Incoming		
	Pref.	**No. of Outgoing**	**Ratio (%)**	**Pref.**	**No. of Incoming**	**Ratio (%)**
1	Tokyo	5,468	12.9	Tokyo	5,987	14.2
2	Osaka	3,347	7.9	Osaka	3,347	7.9
3	Kanagawa	2,475	5.9	Kanagawa	2,418	5.7
4	Aichi	2,409	5.7	Aichi	2,360	5.6
5	Fukuoka	2,063	4.9	Fukuoka	2,052	4.9
6	Saitama	2,009	4.8	Saitama	1,929	4.6
7	Chiba	1,871	4.4	Chiba	1,865	4.4
8	Hyogo	1,677	4.0	Hyogo	1,581	3.7
9	Hokkaido	1,594	3.8	Hokkaido	1,558	3.7
10	Shizuoka	1,128	2.7	Shizuoka	1,115	2.6
11	Hiroshima	977	2.3	Hiroshima	967	2.3
12	Ibaraki	966	2.3	Ibaraki	946	2.2
13	Kyoto	840	2.0	Kyoto	828	2.0
14	Miyagi	758	1.8	Miyagi	751	1.8
15	Okayama	698	1.7	Okayama	689	1.6
16	Kumamoto	692	1.6	Kumamoto	685	1.6
17	Okinawa	664	1.6	Kagoshima	657	1.6
18	Kagoshima	658	1.6	Okinawa	648	1.5
19	Mie	645	1.5	Nagano	639	1.5
20	Tochigi	643	1.5	Mie	635	1.5
21	Nagano	638	1.5	Tochigi	635	1.5
22	Niigata	632	1.5	Niigata	628	1.5
23	Gifu	619	1.5	Gunma	611	1.4
24	Gunma	617	1.5	Fukushima	609	1.4
25	Fukushima	611	1.4	Gifu	609	1.4
26	Ehime	485	1.1	Ehime	482	1.1
27	Nagasaki	475	1.1	Nagasaki	473	1.1
28	Oita	449	1.1	Oita	449	1.1
29	Yamaguchi	440	1.0	Yamaguchi	438	1.0
30	Shiga	430	1.0	Shiga	423	1.0
31	Miyazaki	412	1.0	Miyazaki	412	1.0
32	Nara	400	0.9	Nara	392	0.9
33	Kagawa	358	0.8	Kagawa	360	0.9
34	Wakayama	357	0.8	Ishikawa	356	0.8
35	Ishikawa	356	0.8	Wakayama	354	0.8
36	Aomori	332	0.8	Aomori	333	0.8
37	Iwate	328	0.8	Iwate	327	0.8
38	Yamagata	327	0.8	Yamagata	325	0.8
39	Saga	318	0.8	Yamanashi	312	0.7
40	Yamanashi	314	0.7	Saga	304	0.7
41	Toyama	296	0.7	Toyama	296	0.7
42	Kochi	278	0.7	Kochi	276	0.7
43	Akita	275	0.7	Akita	274	0.6
44	Tokushima	274	0.6	Tokushima	271	0.6
45	Fukui	261	0.6	Fukui	259	0.6
46	Shimane	215	0.5	Shimane	214	0.5
47	Tottori	183	0.4	Tottori	184	0.4
	Total	**42,262**	**100.0**	**Total**	**42,262**	**100.0**

Note: Compiled from data on calls to and from mobile and PHS phones.
*Compiled by TCA based on data publicized by the Ministry of Internal Affairs and Communications

2-2-4-3-2 Main Destination Prefectures by Originating Prefecture (FY2021)

Outgoing	Total number of outgoing calls (million)	Incoming									
		1		2		3		4		5	
		Pref.	Ratio (%)	Pref.	Ratio (%)	Pref.	Ratio (%)	Pref.	Ratio (%)	Pref.	Ratio (%)
Hokkaido	1,594	Hokkaido	91.2	Tokyo	3.5	Kanagawa	0.6	Saitama	0.5	Osaka	0.5
Aomori	332	Aomori	87.2	Tokyo	2.9	Iwate	2.0	Miyagi	2.0	Hokkaido	0.9
Iwate	328	Iwate	83.4	Miyagi	4.7	Tokyo	3.0	Aomori	2.1	Akita	1.2
Miyagi	758	Miyagi	81.6	Tokyo	4.4	Fukushima	2.5	Iwate	2.0	Yamagata	1.5
Akita	275	Akita	86.7	Tokyo	3.2	Miyagi	2.2	Iwate	1.4	Aomori	1.1
Yamagata	327	Yamagata	86.1	Miyagi	3.7	Tokyo	3.1	Fukushima	1.1	Kanagawa	0.7
Fukushima	611	Fukushima	84.5	Tokyo	3.8	Miyagi	3.3	Ibaraki	1.2	Saitama	0.9
Ibaraku	966	Ibaraki	78.7	Tokyo	6.1	Chiba	4.4	Saitama	2.5	Tochigi	2.3
Tochigi	643	Tochigi	78.6	Tokyo	5.4	Ibaraki	3.5	Saitama	3.0	Gunma	2.8
Gunma	617	Gunma	79.8	Saitama	5.2	Tokyo	5.0	Tochigi	3.0	Kanagawa	1.0
Saitama	2,009	Saitama	68.0	Tokyo	17.4	Chiba	2.9	Kanagawa	2.2	Gunma	1.6
Chiba	1,871	Chiba	73.5	Tokyo	13.8	Saitama	2.7	Kanagawa	2.1	Ibaraki	2.1
Tokyo	5,468	Tokyo	71.9	Kanagawa	6.3	Saitama	5.2	Chiba	3.9	Osaka	1.7
Kanagawa	2,475	Kanagawa	72.8	Tokyo	15.8	Chiba	1.8	Saitama	1.7	Shizuoka	1.0
Niigata	632	Niigata	87.6	Tokyo	3.8	Saitama	1.0	Nagano	0.8	Kanagawa	0.8
Toyama	296	Toyama	84.5	Ishikawa	4.1	Tokyo	2.8	Osaka	1.3	Aichi	1.1
Ishikawa	356	Ishikawa	83.6	Toyama	3.2	Tokyo	2.8	Fukui	1.9	Osaka	1.7
Fukui	261	Fukui	84.7	Ishikawa	2.9	Tokyo	2.2	Osaka	2.1	Aichi	1.2
Yamanashi	314	Yamanashi	83.2	Tokyo	6.2	Kanagawa	2.2	Nagano	1.6	Shizuoka	1.5
Nagano	638	Nagano	85.9	Tokyo	4.2	Aichi	1.2	Saitama	1.0	Kanagawa	0.9
Gifu	619	Gifu	77.7	Aichi	11.7	Tokyo	2.6	Osaka	1.2	Mie	0.9
Shizuoka	1,128	Shizuoka	84.4	Tokyo	4.4	Aichi	3.0	Kanagawa	2.0	Osaka	0.9
Aichi	2,409	Aichi	82.8	Tokyo	4.2	Gifu	3.0	Mie	1.6	Osaka	1.6
Mie	645	Mie	81.8	Aichi	6.6	Tokyo	2.3	Osaka	2.0	Gifu	0.9
Shiga	430	Shiga	75.9	Kyoto	6.2	Osaka	5.7	Tokyo	2.5	Aichi	1.4
Kyoto	840	Kyoto	75.0	Osaka	8.8	Tokyo	3.3	Shiga	3.1	Hyogo	2.2
Osaka	3,347	Osaka	77.1	Tokyo	4.9	Hyogo	4.7	Kyoto	2.1	Nara	1.5
Hyogo	1,677	Hyogo	75.1	Osaka	12.7	Tokyo	3.3	Kyoto	1.2	Chiba	0.8
Nara	400	Nara	71.8	Osaka	13.1	Tokyo	3.8	Kyoto	2.9	Hyogo	1.6
Wakayama	357	Wakayama	83.4	Osaka	7.9	Tokyo	2.0	Nara	1.1	Hyogo	1.0
Tottori	183	Tottori	83.2	Shimane	4.4	Okayama	1.9	Tokyo	1.9	Osaka	1.7
Shimane	215	Shimane	83.5	Tottori	3.8	Hiroshima	3.6	Tokyo	1.7	Osaka	1.4
Okayama	698	Okayama	84.1	Hiroshima	3.6	Tokyo	2.3	Osaka	1.9	Hyogo	1.5
Hiroshima	977	Hiroshima	83.9	Tokyo	2.8	Okayama	2.4	Yamaguchi	1.9	Osaka	1.7
Yamaguchi	440	Yamaguchi	83.2	Hiroshima	4.2	Fukuoka	3.8	Tokyo	2.1	Osaka	1.2
Tokushima	274	Tokushima	86.1	Kagawa	3.0	Osaka	1.9	Tokyo	1.8	Hyogo	1.3
Kagawa	358	Kagawa	82.7	Ehime	2.6	Tokyo	2.3	Tokushima	2.1	Osaka	2.0
Ehime	485	Ehime	86.3	Tokyo	2.3	Kagawa	2.2	Osaka	1.5	Hiroshima	1.3
Kochi	278	Kochi	88.6	Tokyo	1.8	Ehime	1.8	Kagawa	1.6	Osaka	1.4
Fukuoka	2,063	Fukuoka	83.8	Tokyo	2.9	Saga	1.9	Kumamoto	1.6	Oita	1.4
Saga	318	Saga	73.4	Fukuoka	15.7	Nagasaki	3.1	Tokyo	1.8	Kumamoto	1.0
Nagasaki	475	Nagasaki	86.2	Fukuoka	4.5	Saga	2.0	Tokyo	1.9	Osaka	0.8
Kumamoto	692	Kumamoto	85.7	Fukuoka	5.0	Tokyo	1.9	Kagoshima	1.2	Osaka	0.8
Oita	449	Oita	85.6	Fukuoka	5.8	Tokyo	1.8	Kumamoto	1.0	Osaka	0.8
Miyazaki	412	Miyazaki	86.7	Kagoshima	3.0	Fukuoka	2.4	Tokyo	1.9	Kumamoto	1.3
Kagoshima	658	Kagoshima	88.0	Fukuoka	2.3	Miyazaki	2.1	Tokyo	1.9	Kumamoto	1.2
Okinawa	664	Okinawa	91.6	Tokyo	3.2	Fukuoka	0.9	Osaka	0.8	Kanagawa	0.4

Note: Compiled from data on calls to and from mobile and PHS phones.
*Compiled by TCA based on data publicized by the Ministry of Internal Affairs and Communications

2-2-4-3-3 Main Originating Prefectures by Destination Prefecture (FY2021)

Incoming	Total number of Incoming calls (million)	Outgoing									
		1		2		3		4		5	
		Pref.	Ratio (%)	Pref.	Ratio (%)	Pref.	Ratio (%)	Pref.	Ratio (%)	Pref.	Ratio (%)
Hokkaido	1,558	Hokkaido	93.3	Tokyo	2.1	Kanagawa	0.5	Osaka	0.5	Saitama	0.4
Aomori	333	Aomori	87.1	Tokyo	2.3	Iwate	2.1	Miyagi	1.9	Akita	0.9
Iwate	327	Iwate	83.6	Miyagi	4.6	Tokyo	2.3	Aomori	2.1	Akita	1.2
Miyagi	751	Miyagi	82.3	Tokyo	3.0	Fukushima	2.7	Iwate	2.0	Yamagata	1.6
Akita	274	Akita	87.2	Tokyo	2.4	Miyagi	2.0	Iwate	1.4	Aomori	1.0
Yamagata	325	Yamagata	86.6	Miyagi	3.5	Tokyo	2.3	Fukushima	1.1	Kanagawa	0.8
Fukushima	609	Fukushima	84.7	Miyagi	3.2	Tokyo	2.7	Ibaraki	1.3	Saitama	1.0
Ibaraki	946	Ibaraki	80.3	Tokyo	4.7	Chiba	4.2	Saitama	2.6	Tochigi	2.4
Tochigi	635	Tochigi	79.7	Tokyo	3.9	Ibaraki	3.5	Saitama	3.2	Gunma	3.0
Gunma	611	Gunma	80.5	Saitama	5.4	Tokyo	4.0	Tochigi	2.9	Kanagawa	1.1
Saitama	1,929	Saitama	70.8	Tokyo	14.8	Chiba	2.6	Kanagawa	2.2	Gunma	1.7
Chiba	1,865	Chiba	73.7	Tokyo	11.5	Saitama	3.1	Kanagawa	2.3	Ibaraki	2.3
Tokyo	5,987	Tokyo	65.6	Kanagawa	6.5	Saitama	5.8	Chiba	4.3	Osaka	2.7
Kanagawa	2,418	Kanagawa	74.6	Tokyo	14.2	Saitama	1.8	Chiba	1.7	Shizuoka	1.0
Niigata	628	Niigata	88.1	Tokyo	3.0	Saitama	1.1	Kanagawa	0.8	Nagano	0.8
Toyama	296	Toyama	84.4	Ishikawa	3.9	Tokyo	2.5	Osaka	1.2	Aichi	1.1
Ishikawa	356	Ishikawa	83.6	Toyama	3.4	Tokyo	2.2	Fukui	2.1	Osaka	1.5
Fukui	259	Fukui	85.4	Ishikawa	2.7	Osaka	1.9	Tokyo	1.8	Aichi	1.2
Yamanashi	312	Yamanashi	83.7	Tokyo	5.4	Kanagawa	2.3	Nagano	1.5	Shizuoka	1.5
Nagano	639	Nagano	85.8	Tokyo	3.6	Aichi	1.3	Saitama	1.1	Kanagawa	1.0
Gifu	609	Gifu	79.0	Aichi	11.7	Tokyo	1.7	Osaka	1.1	Mie	1.0
Shizuoka	1,115	Shizuoka	85.4	Tokyo	3.2	Aichi	2.9	Kanagawa	2.2	Osaka	1.0
Aichi	2,360	Aichi	84.6	Gifu	3.1	Tokyo	2.3	Mie	1.8	Shizuoka	1.4
Mie	635	Mie	83.1	Aichi	6.3	Osaka	2.1	Tokyo	1.5	Gifu	0.9
Shiga	423	Shiga	77.1	Kyoto	6.1	Osaka	5.4	Tokyo	1.6	Hyogo	1.5
Kyoto	828	Kyoto	76.1	Osaka	8.6	Shiga	3.2	Hyogo	2.4	Tokyo	2.0
Osaka	3,347	Osaka	77.1	Hyogo	6.3	Tokyo	2.7	Kyoto	2.2	Nara	1.6
Hyogo	1,581	Hyogo	79.7	Osaka	9.9	Tokyo	1.9	Kyoto	1.2	Okayama	0.7
Nara	392	Nara	73.4	Osaka	13.2	Kyoto	3.0	Hyogo	1.8	Tokyo	1.6
Wakayama	354	Wakayama	84.2	Osaka	7.6	Tokyo	1.2	Hyogo	1.2	Nara	1.1
Tottori	184	Tottori	83.0	Shimane	4.4	Okayama	2.0	Osaka	1.8	Hyogo	1.7
Shimane	214	Shimane	83.8	Tottori	3.8	Hiroshima	3.4	Tokyo	1.5	Osaka	1.4
Okayama	689	Okayama	85.1	Hiroshima	3.4	Osaka	2.0	Tokyo	1.6	Hyogo	1.5
Hiroshima	967	Hiroshima	84.8	Okayama	2.6	Yamaguchi	1.9	Tokyo	1.7	Osaka	1.6
Yamaguchi	438	Yamaguchi	83.7	Hiroshima	4.3	Fukuoka	3.8	Tokyo	1.4	Osaka	1.1
Tokushima	271	Tokushima	87.0	Kagawa	2.7	Osaka	2.0	Hyogo	1.4	Tokyo	1.3
Kagawa	360	Kagawa	82.4	Ehime	2.9	Tokushima	2.3	Osaka	2.0	Tokyo	1.7
Ehime	482	Ehime	86.7	Tokyo	2.1	Kagawa	1.9	Osaka	1.5	Hiroshima	1.3
Kochi	276	Kochi	89.2	Ehime	1.8	Osaka	1.4	Kagawa	1.4	Tokyo	1.3
Fukuoka	2,052	Fukuoka	84.3	Saga	2.4	Tokyo	1.9	Kumamoto	1.7	Oita	1.3
Saga	304	Saga	76.6	Fukuoka	13.0	Nagasaki	3.1	Tokyo	1.4	Kumamoto	1.0
Nagasaki	473	Nagasaki	86.6	Fukuoka	4.3	Saga	2.1	Tokyo	1.5	Kumamoto	0.8
Kumamoto	685	Kumamoto	86.5	Fukuoka	4.9	Tokyo	1.4	Kagoshima	1.2	Miyazaki	0.8
Oita	449	Oita	85.6	Fukuoka	6.2	Tokyo	1.3	Kumamoto	1.1	Miyazaki	0.7
Miyazaki	412	Miyazaki	86.7	Kagoshima	3.3	Fukuoka	2.3	Tokyo	1.5	Kumamoto	1.3
Kagoshima	657	Kagoshima	88.2	Fukuoka	2.2	Miyazaki	1.9	Tokyo	1.6	Kumamoto	1.2
Okinawa	648	Okinawa	93.8	Tokyo	1.7	Fukuoka	0.8	Osaka	0.5	Kanagawa	0.4

Note: Compiled from data on calls to and from mobile and PHS phones.
*Compiled by TCA based on data publicized by the Ministry of Internal Affairs and Communications

2-2-5 Situation of Traffic of International Telephone Services

2-2-5-1 Trends in Number and Duration of International Telephone Calls

(Million calls, Million minutes)

Category		FY2017	FY2018	FY2019	FY2020	FY2021
Number of Calls	Outgoing	194.8	159.1	137.9	50.0	36.4
	Incoming	298.6	289.3	333.5	317.6	462.0
	Total	**493.4**	**448.5**	**471.4**	**367.6**	**498.5**
Duration of calls	Outgoing	744.4	594.3	496.5	258.5	174.2
	Incoming	902.1	750.9	661.1	527.1	520.9
	Total	**1,646.5**	**1,345.2**	**1,157.6**	**785.7**	**695.2**

*Compiled by TCA based on data publicized by the Ministry of Internal Affairs and Communications

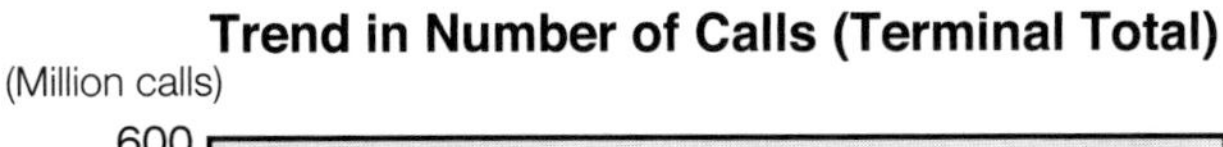

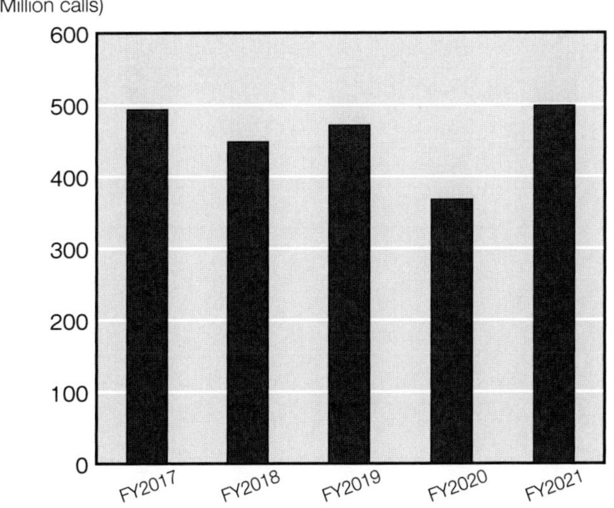

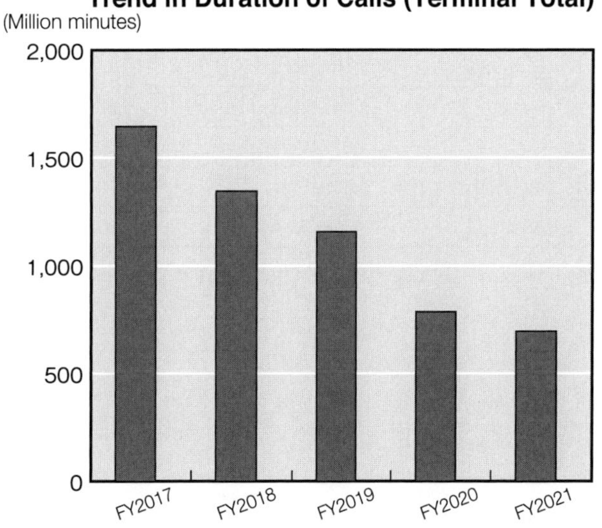

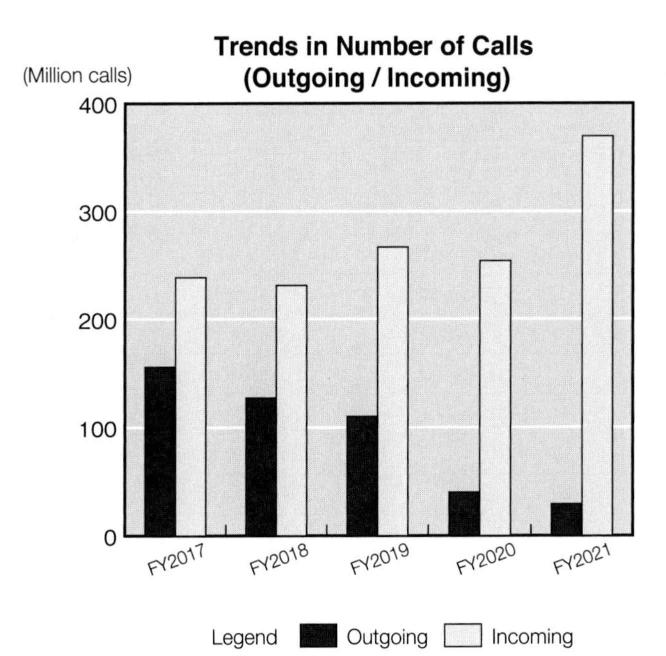

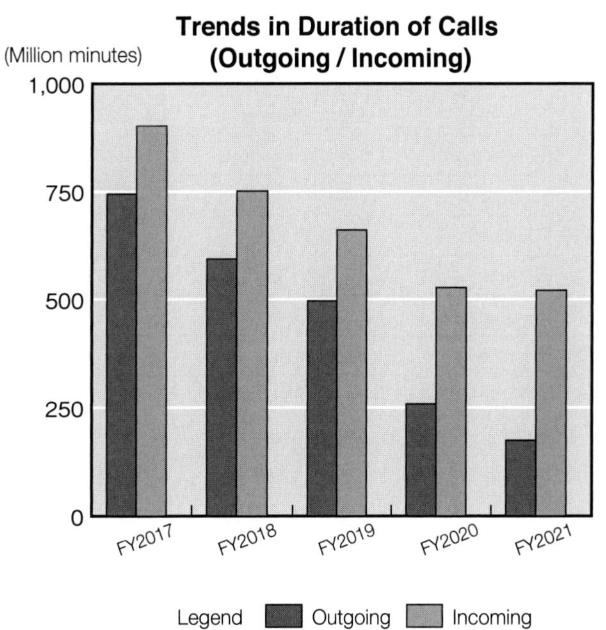

2-2-5-2 Situation of Duration of International Calls by Country/Region (Top Countries/Regions Shown)

2-2-5-2-1 Trends in Share of Outgoing Call Duration by Country/Region

Ranking	FY2017		FY2018		FY2019		FY2020		FY2021	
1	China	20.93%	U.S.A. (mainland)	19.33%	U.S.A. (mainland)	19.83%	U.S.A. (mainland)	35.13%	U.S.A. (mainland)	33.04%
2	U.S.A. (mainland)	17.79%	China	17.75%	Hong Kong	19.19%	China	16.15%	China	17.07%
3	Hong Kong	10.80%	Hong Kong	15.84%	China	16.46%	Hong Kong	8.86%	Korea	7.09%
4	Philippines	8.46%	Philippines	6.36%	Korea	5.16%	Korea	6.26%	Hong Kong	4.53%
5	Korea	6.01%	Korea	6.06%	Thailand	3.49%	Thailand	3.51%	Philippines	3.77%
6	Thailand	3.63%	Thailand	3.74%	Philippines	3.34%	Philippines	3.49%	Thailand	3.77%
7	Taiwan	3.11%	Taiwan	3.19%	Taiwan	3.02%	Taiwan	3.20%	Bangladesh	3.38%
8	Singapore	2.83%	Singapore	2.80%	Singapore	2.85%	Singapore	2.97%	Taiwan	3.36%
9	India	2.34%	India	2.49%	India	2.69%	U.K.	2.01%	Singapore	2.46%
10	Vietnam	1.76%	Germany	1.80%	U.K.	2.01%	India	1.71%	U.K.	2.02%
11	Germany	1.68%	U.K.	1.74%	Germany	1.98%	Germany	1.68%	Germany	1.58%
12	Macau	1.64%	Macau	1.68%	Bangladesh	1.61%	France	1.30%	Vietnam	1.41%
13	U.K.	1.61%	Vietnam	1.50%	Australia	1.60%	Vietnam	1.17%	France	1.40%
14	Indonesia	1.53%	France	1.42%	France	1.56%	Indonesia	1.13%	India	1.35%
15	Australia	1.39%	Australia	1.31%	Macau	1.47%	Australia	1.10%	Australia	1.19%

*Compiled by TCA based on data publicized by the Ministry of Internal Affairs and Communications

2-2-5-2-2 Trends in Share of Incoming Call Duration by Country/Region

Ranking	FY2017		FY2018		FY2019		FY2020		FY2021	
1	U.S.A. (mainland)	18.75%	China	22.43%	China	25.12%	U.S.A. (mainland)	27.52%	China	35.60%
2	China	18.50%	U.S.A. (mainland)	20.30%	U.S.A. (mainland)	20.12%	Korea	27.40%	Korea	29.32%
3	Korea	12.60%	Korea	18.48%	Korea	18.92%	China	26.51%	U.S.A. (mainland)	25.90%
4	Hong Kong	8.82%	Hong Kong	12.73%	Hong Kong	14.03%	Hong Kong	3.17%	Hong Kong	0.93%
5	Taiwan	4.26%	Canada	2.33%	Canada	3.16%	Canada	2.05%	Australia	0.86%
6	Luxembourg	3.29%	Singapore	2.14%	Singapore	2.45%	Australia	1.62%	U.K.	0.81%
7	Germany	2.87%	Luxembourg	1.75%	Taiwan	1.23%	Singapore	1.57%	Singapore	0.76%
8	Thailand	2.83%	France	1.73%	Australia	1.20%	Germany	1.38%	Taiwan	0.74%
9	France	2.70%	Taiwan	1.70%	Germany	1.15%	Thailand	0.96%	Thailand	0.53%
10	Singapore	2.69%	Germany	1.66%	Macau	1.08%	Taiwan	0.91%	UAE	0.50%
11	Canada	2.68%	Malaysia	1.48%	Malaysia	1.06%	Malaysia	0.86%	Indonesia	0.47%
12	U.K.	2.12%	Thailand	1.47%	Thailand	1.05%	Belgium	0.77%	Malaysia	0.44%
13	Belgium	1.95%	Macau	1.30%	France	0.89%	U.K.	0.66%	Belgium	0.43%
14	Indonesia	1.90%	Indonesia	1.24%	Iceland	0.77%	UAE	0.54%	Germany	0.38%
15	Malaysia	1.58%	Australia	1.11%	Indonesia	0.74%	Vietnam	0.53%	Vietnam	0.34%

*Compiled by TCA based on data publicized by the Ministry of Internal Affairs and Communications

2-2-5-2-3 Outgoing and Incoming Call Duration by Country/Region (FY2021)

Country/Region (descending order according to outgoing duration)	Outgoing from Japan						Incoming to Japan					
	Ranking in outgoing		Duration of outgoing (Million minutes)	Increase or decrease ratio over previous year (%)	Share (%)	Accumu lated share (%)	Ranking in incoming		Duration of incoming (Million minutes)	Increase or decrease ratio over previous year (%)	Share (%)	Accumu lated share (%)
	2021	2020					2021	2020				
U.S.A. (mainland)	1	(1)	57.6	▲41.52%	33.04%	33.04%	3	(1)	134.9	1.43%	25.90%	25.90%
China	2	(2)	29.7	▲63.61%	17.07%	50.11%	1	(3)	185.4	11.68%	35.60%	61.50%
Korea	3	(4)	12.3	▲51.83%	7.09%	57.20%	2	(2)	152.8	22.12%	29.32%	90.82%
Hong Kong	4	(3)	7.9	▲91.72%	4.53%	61.72%	4	(4)	4.8	▲94.78%	0.93%	91.75%
Philippines	5	(6)	6.6	▲60.39%	3.77%	65.50%	17	(18)	1.1	▲55.18%	0.20%	91.95%
Thailand	6	(5)	6.6	▲62.14%	3.77%	69.26%	9	(9)	2.7	▲60.47%	0.53%	92.48%
Bangladesh	7	(34)	5.9	▲26.19%	3.38%	72.65%	47	(49)	0.0	▲76.25%	0.01%	92.49%
Taiwan	8	(7)	5.9	▲60.91%	3.36%	76.01%	8	(10)	3.8	▲52.75%	0.74%	93.22%
Singapore	9	(8)	4.3	▲69.74%	2.46%	78.46%	7	(7)	4.0	▲75.62%	0.76%	93.98%
U.K	10	(9)	3.5	▲64.64%	2.02%	80.49%	6	(13)	4.2	▲6.46%	0.81%	94.79%
Germany	11	(11)	2.7	▲72.05%	1.58%	82.06%	14	(8)	2.0	▲73.81%	0.38%	95.17%
Vietnam	12	(13)	2.5	▲64.41%	1.41%	83.47%	15	(15)	1.8	▲59.70%	0.34%	95.51%
France	13	(12)	2.4	▲68.52%	1.40%	84.87%	16	(17)	1.4	▲76.46%	0.27%	95.78%
India	14	(10)	2.3	▲82.39%	1.35%	86.22%	24	(20)	0.4	▲88.14%	0.08%	95.86%
Australia	15	(15)	2.1	▲73.94%	1.19%	87.41%	5	(6)	4.5	▲43.49%	0.86%	96.72%
Hawaii (U.S.A.)	16	(18)	2.0	▲51.44%	1.13%	88.54%	22	(19)	0.5	▲66.76%	0.10%	96.83%
Indonesia	17	(14)	2.0	▲68.39%	1.13%	89.67%	11	(16)	2.4	▲50.32%	0.47%	97.29%
Malaysia	18	(16)	1.9	▲59.13%	1.07%	90.74%	12	(11)	2.3	▲67.42%	0.44%	97.73%
Canada	19	(17)	1.7	▲53.79%	0.97%	91.71%	18	(5)	0.7	▲96.44%	0.14%	97.88%
Belgium	20	(19)	1.1	▲18.95%	0.65%	92.37%	13	(12)	2.2	▲37.37%	0.43%	98.30%
Italy	21	(22)	0.7	▲76.80%	0.39%	92.76%	26	(27)	0.3	▲61.43%	0.06%	98.36%
Cuba	22	(91)	0.6	648.40%	0.36%	93.13%	76	(111)	0.0	1.59%	0.00%	98.37%
Brazil	23	(21)	0.6	▲76.07%	0.35%	93.47%	27	(26)	0.3	▲61.80%	0.06%	98.42%
UAE	24	(26)	0.5	▲65.95%	0.31%	93.79%	10	(14)	2.6	▲30.52%	0.50%	98.92%
Myanmar	25	(30)	0.5	▲45.27%	0.31%	94.09%	19	(29)	0.7	57.34%	0.14%	99.06%
Sri Lanka	26	(27)	0.5	▲64.35%	0.30%	94.40%	20	(23)	0.6	▲48.46%	0.12%	99.18%
Belarus	27	(158)	0.5	3859.35%	0.30%	94.69%	153	(128)	0.0	▲98.64%	0.00%	99.18%
Netherlands	28	(23)	0.5	▲71.19%	0.29%	94.99%	33	(31)	0.1	▲64.15%	0.03%	99.21%
Satellite phones	29	(33)	0.5	▲14.10%	0.26%	95.25%	23	(21)	0.4	▲59.64%	0.08%	99.29%
Mexico	30	(28)	0.5	▲59.87%	0.26%	95.51%	21	(22)	0.6	▲52.95%	0.11%	99.40%
Total of other countries	—	—	7.8		4.49%	100.00%	—	—	3.1		0.60%	100.00%
Total of all countries/regions	—	—	**174.2**		—	—	—	—	**520.9**		—	—

*Compiled by TCA based on data publicized by the Ministry of Internal Affairs and Communications

2-3 Movements of Services and Charges

2-3-1 Fixed Telephones

2-3-1-1 Progress of Rates

2-3-1-1-1 Progress of Telephone Rates of NTT

1985	A three-minute call to the longest distance zone covering over 320km cost ¥400.
July 1986	First reduction of rates after NTT privatization was implemented. The Saturday discount was introduced, which applied, as was the case with holidays and nighttime, 40% discount from the normal rates for weekdays.
February 1988	NTT reduced the longest distance rate for weekday daytime calls to a level of ¥360 for 3 minutes.
February 1989	NTT reduced the longest distance rate for weekday daytime calls to a level of ¥330 for 3 minutes. It also cut rates for calls to the adjacent distance zone and areas within a radius of 20km from a level of ¥30 to ¥20 for 3 minutes (First reduction for short-distance calls since 1972).
March 1990	NTT reduced the longest distance rate for weekday daytime calls to a level of ¥280 for 3 minutes. It also introduced late-night discounts for local, short- and middle-distance calls.
March 1991	Distance segments covering over 160km were consolidated into a single longest distance zone, and the longest distance rate for weekday daytime calls was reduced to a level of ¥240 for 3 minutes. NTT also reduced rates for weekday daytime calls to areas within 20-30km radiuses to a level of ¥40 for 3 minutes. Late-night discount time period was extended by two hours to cover from 11 p.m. to 8 a.m. in the next morning.
June 1992	NTT reduced the longest distance rate for weekday daytime calls to a level of ¥200 for 3 minutes.
October 1993	NTT streamlined the distance segments covering 30-100km to two from four steps, and reduced rates for portions exceeding 30km by ¥10-60. The longest distance rate for weekday daytime calls was reduced to a level of ¥180 for 3 minutes.
March 1996	The rate for longest distance calls was lowered to ¥140 per 3 minutes in the daytime on weekdays.
February 1997	The rate for long-distance calls over 100 km was lowered to ¥110 per 3 minutes in the daytime on weekdays.
February 1998	Distances of over 100km were incorporated into the longest distance rate zone, and the longest distance rate for weekday daytime calls was reduced to a level of ¥90 for 3 minutes.
July 1999	With the reorganization of NTT, NTT East and NTT West took charge of intra-prefecture calls, and NTT Communications took inter-prefectures calls.
October 2000	NTT East and West lowered the toll call rate over 20 km in distance. The rate per 3 minutes in the daytime on weekdays was lowered to ¥30 for 20 to 60 km, and to ¥40 for over 60 km.
January 2001	NTT East reduced the local call rates to ¥9 per 3 minutes.
May 2001	NTT East and West lowered the local call rate to ¥8.5 per 3 minutes both in the daytime and at night.

2-3-1-1-2 Progress of Rates of Long-Distance and International NCCs

September 1987	Three new long-distance carriers stated services. DDI CORPORATION, JAPAN TELECOM CO., LTD. and Teleway Japan Corporation started services. They offer charges about 25% below those of NTT. A 3-minute weekday daytime call to the longest distance zone of 340km cost ¥300 (in the case of NTT-established local portion charge being ¥20).
February 1988	These NCCs reduced evening and late-night rates, and introduced evening discounts into short-distance rates.
February 1989	Rates applicable to all the distance zones were reduced, bringing the longest distance rate for weekday daytime calls down to a level of ¥280 for 3 minutes.
March 1990	The longest distance rate for weekday daytime calls was reduced to a level of ¥240 for 3 minutes. Rates for calls to all the distance zones for evening, Saturdays, Sundays and holidays were reduced.
March 1991	Distance zones covering over 170km were consolidated into the longest distance zone, and the longest distance rate for weekday daytime calls was reduced to a level of ¥200 for 3 minutes. Evening, Saturday, Sunday, and holiday rates were also reduced.
April 1992	The longest distance rate for weekday daytime calls was reduced to a level of ¥180 for 3 minutes.
November 1993	In response to the introduction of the end-to-end charging (that was established by NCCs on an end-to-end basis for the entirety from the calling party through the called party including the local portion) in place of the add-on charging so far applied (total of the charge for trunk portion established by NCCs, and the charge for local portion established by NTT), an overall reduction of rates was implemented. As a result, the longest distance rate for weekday daytime calls was reduced to a level of ¥170 for 3 minutes. The late-night discount time zone (from 11 p.m. to 8 a.m. in the next morning) was established, and the distance zones covering from 60km up to 100km were combined from two to one.
April 1994	The charge for the end portion provided by NTT was changed from the user charge to the cost-based inter-carriers settlement charge (access charge).
March 1996	In response to the reduction of the inter-carrier settlement charges paid by NCCs to NTT relating to the local portion provided by NTT (so-called "access charge"), the longest distance (over 170km) rate for weekday daytime calls was reduced to a level of ¥130 for 3 minutes from ¥170. In addition, the distance zone for short-distance calls, which had been set up in terms of "up to 60km" was divided into two zones, "up to 30km" and "over 30km up to 60km", and the rate-cut was made for "up to 30km" weekday daytime calls, and "up to 30km" and "over 30km up to 60km" late-night and early morning calls.
February 1997	The longest distance rate for weekday daytime calls was reduced to a level of ¥100 for 3 minutes.
February 1998	The longest distance rate for weekday daytime calls was reduced to a level of ¥90 for 3 minutes (The reduction brought NCCs' rates to the same level as NTT's). The distance zones for the adjacent zone and the inside radius of 20km were established.
July 1998	KDD made a full-scaled inroad into domestic telephone markets, setting the longest distance rate for weekday daytime calls at a level of ¥69 for 3 minutes.
April 2000	Daytime and evening rates, etc to 20 - 30km and 30 - 60km distance zones were reduced NTT Communications reduced daytime and evening rates for calls to 30 - 60km and 60 - 100km distance zones, and evening and midnight rates for 60 - 100km and over 100km distance zones.
October 2000	KDD, DDI and IDO merged into KDDI. New Intra-prefecture rates were established at a level of ¥40 for 3-minute weekday daytime call to the 60km or longer distance zone.

December 2000 C&W IDC fully entered the local domestic telephone market, and started the service setting, at a level of ¥45, its remotest distance rate applicable to 3-minute calls of 100km or longer distances for all day.

March 2001 The rate to the remotest distance zone was reduced to a level of ¥80 for 3-minute weekday daytime call, and the rate applicable to the 60-100km distance zone to a level of ¥60 for 3-minute weekday daytime call.
NTT Communications reduced rates applicable to the 20 - 30km distance zone for all day, the 30 - 60km distance zone during evening and midnight, the 60 - 100km distance zone during midnight, and the more than 100km distance zone during daytime and midnight.

April 2001 Fusion Communications started IP telephone service, establishing its rate at ¥20 for 3-minute irrespective of distance throughout Japan.

May 2001 NTT Communications entered the local call market in Tokyo, Aichi, and Osaka. The rate is ¥8.5 per 3 minutes.
KDDI and Japan Telecom entered the local call market. Their local call rate is ¥8.5 for 3-minute weekday daytime call.

December 2004 Japan Telecom started "OTOKU Line" fixed telephone service.

February 2005 KDDI started "Metal Plus" telephone service.

June 2006 Japan Telecom Co. Ltd. took over telecommunications business from Heisei Denden Corp. and Heisei Denden Communications Corp.

October 2006 Japan Telecom Co. Ltd. changed its company name to SoftBank Telecom Corp.

April 2015 SoftBank Mobile Corp., SoftBank BB Corp., SoftBank Telecom Corp., and Ymobile Corporation merged together to form SoftBank Mobile Corp.

July 2015 SoftBank Mobile Corp. changed its company name to SoftBank Corp.

December 2015 Fusion Communications Corp. changed its company name to Rakuten Communications Corp.

June 2016 KDDI terminated its "Metal Plus" telephone service.

July 2019 Rakuten Communications Corp. transferred its domestic telephone service (MYLINE) and the Rakuten Denwa phone service to Rakuten Mobile, Inc. through a company split.

(Reference) Carriers Participating in MYLINE

(As of October 2023)

Carrier \ Call Catogory	ID number of telephone company	Local	Intra-pref long-distance	Outside of Prefecture	International	Registration available in
NTT East CORPORATION	0036	○	○			Eastern Japan
NTT West CORPORATION	0039	○	○			Western Japan
NTT Communications Corporation	0033	○	○	○	○	Nationwide
KDDI CORPORATION	0077 001 (International call)	○	○	○	○	Nationwide
SoftBank Corp.	0088 0061 (International call)	○	○	○	○	Nationwide
Rakuten Mobile, Inc.	0038	○	○	○	○	Nationwide
ARTERIA Networks Corporation	0060	○	○	○	○	Tokyo and 18 prefectures

*MY LINE Website : http://www.myline.org/index_e.html

2-3-1-1-3 Progress of Rates of Regional and Cable TV Operators

May 1988 Tokyo Telecommunication Network Company Inc. (called TTNet hereafter, later reformed to the present Poweredcom), a regional common carrier, started direct subscriber telephone service.

June 1997 Cable TV operator, TITUS COMMUNICATIONS CORPORATION, started subscriber telephone services. For call billing the Hudson charging method in units of 20 seconds was introduced.

July 1997	Suginami Cable TV Co., Ltd. (currently J-COM Tokyo) started subscriber telephone services.
January 1998	TTNet started relay telephone services with the rate of ¥9 for 3 minutes intra-zone calls, and the longest distance rate set at ¥72 for 3 minutes on weekday daytime calls.
March 1998	TTNet reduced the longest distance rate for weekday daytime calls to a level of ¥63 for 3 minutes.
April 1999	Kyushu Telecommunication Network Co., LTD. (hereafter, QTNet) started relay telephone services with the rate of ¥9 for intra-zone calls for 3 minutes on weekdays during the daytime, and ¥70 for the longest distance.
May 2000	TTNet reduced the rate for 3-minute weekday daytime call to 60 - 100km distance zone from ¥54 to ¥45.
November 2000	QTNet established new intra-prefecture rate, setting weekday daytime rate for call to a 60km or longer distance zone at a level of ¥27 for 3-minute.
May 2001	TTNet reduced the charges for calls to all the distance zones. The charge applicable to the remotest distance zone was reduced to a level of ¥54 for 3-minute daytime call, ¥36 for 3-minute daytime call to a 60 -100km distance zone, and ¥8.4 for local calls, respectively. QTNet reduced the rate for intra-zone calls to ¥8.4 for three minutes during the day on weekdays.
April 2003	POWEREDCOM merged with TTNet, and the new company was named POWEREDCOM, Inc.
July 2004	The telephone business of POWEREDCOM is merged with FUSION COMMUNICATIONS CORP.
June 2018	QTnet (formerly Kyushu Telecommunication Network) terminated its relay telephone services.
April 2019	K-Opticom Corporation changed its company name to OPTAGE Inc.

2-3-1-1-4 Progress of ISDN Service Provision

April 1988	NTT inaugurated ISDN service.
October 1995	Osaka Media Port and Shikoku Information and Telecommunication Network inaugurated ISDN service.
February 1996	NTT started "INS Telehodai", a fixed rate service to selected telephone numbers in the midnight to early morning time zone.
March 1996	HOKKAIDO TELECOMMUNICATION NETWORK and Tohoku Intelligent Telecommunication inaugurated ISDN service.
April 1996	Chubu Telecommunications inaugurated ISDN service.
April 1997	TTNet and QTNet inaugurated ISDN service.
July 1997	NTT inaugurated ISDN service free of the facilities installation charge, "INS Net 64 Lite".
October 1997	Chugoku Telecommunication Network inaugurated ISDN service.
December 1997	Osaka Media Port started interconnection with NTT.
July 2000	NTT East and NTT West inaugurated fixed rate IP connection service, "FLET'S ISDN".
July 2003	Chugoku Telecommunication Network merged with Chugoku Information System Service and reorganized as Energia Communications.
April 2010	Tohoku Intelligent Telecommunication terminated ISDN service.
March 2011	Energia Communications terminated ISDN service.
December 2013	QTNet terminated its ISDN service.

• Changes in NTT's Call Rates (for a 3-minute weekday daytime call)

<table>
<tr><th>Time of Revision</th><th>Number of Distance Zone</th><th>Within Zone</th><th>Adjacent Zone up to 20km</th><th>-30km</th><th>-40km</th><th>-60km</th><th>-80km</th><th>-100km</th><th>-120km</th><th>-160km</th><th>-240km</th><th>-320km</th><th>-500km</th><th>-750km</th><th>Over 750km</th></tr>
<tr><td>Before Aug. 1983</td><td>14</td><td>10</td><td>30</td><td>50</td><td>60</td><td>90</td><td>120</td><td>140</td><td>180</td><td>230</td><td>280</td><td>360</td><td>450</td><td>600</td><td>720</td></tr>
<tr><td>Aug. 1983</td><td>14</td><td>10</td><td>30</td><td>50</td><td>60</td><td>90</td><td>120</td><td>140</td><td>180</td><td>230</td><td>280</td><td>360</td><td>450</td><td>520</td><td>600</td></tr>
<tr><td>Jul. 1985</td><td>12</td><td>10</td><td>30</td><td>50</td><td>60</td><td>90</td><td>120</td><td>140</td><td>180</td><td>230</td><td>280</td><td>360</td><td colspan="3">400</td></tr>
<tr><td>Jul. 1986</td><td>10</td><td>10</td><td>30</td><td>50</td><td>60</td><td>90</td><td>120</td><td>140</td><td colspan="2">180</td><td colspan="2">260</td><td colspan="3">400</td></tr>
<tr><td>Feb. 1988</td><td>10</td><td>10</td><td>30</td><td>50</td><td>60</td><td>90</td><td>120</td><td>140</td><td colspan="2">180</td><td colspan="2">260</td><td colspan="3">360</td></tr>
<tr><td>Feb. 1989</td><td>10</td><td>10</td><td>30</td><td>50</td><td>60</td><td>90</td><td>120</td><td>140</td><td colspan="2">180</td><td colspan="2">260</td><td colspan="3">330</td></tr>
<tr><td>Mar. 1990</td><td>10</td><td>10</td><td>30</td><td>50</td><td>60</td><td>90</td><td>120</td><td>140</td><td colspan="2">180</td><td colspan="2">260</td><td colspan="3">280</td></tr>
<tr><td>Mar. 1991</td><td>9</td><td>10</td><td>30</td><td>40</td><td>60</td><td>90</td><td>120</td><td>140</td><td colspan="2">180</td><td colspan="5">240</td></tr>
<tr><td>Jun. 1992</td><td>9</td><td>10</td><td>30</td><td>40</td><td>60</td><td>90</td><td>120</td><td>140</td><td colspan="2">180</td><td colspan="5">200</td></tr>
<tr><td>Oct. 1993</td><td>7</td><td>10</td><td>30</td><td>40</td><td colspan="2">50</td><td colspan="2">80</td><td colspan="2">140</td><td colspan="5">180</td></tr>
<tr><td>Mar. 1996</td><td>6</td><td>10</td><td>30</td><td>40</td><td colspan="2">50</td><td colspan="2">80</td><td colspan="7">140</td></tr>
<tr><td>Feb. 1997</td><td>6</td><td>10</td><td>30</td><td>40</td><td colspan="2">50</td><td colspan="2">80</td><td colspan="7">110</td></tr>
<tr><td>Feb. 1998</td><td>6</td><td>10</td><td>30</td><td>40</td><td colspan="2">50</td><td colspan="2">80</td><td colspan="7">90</td></tr>
<tr><td>(inter-Pref.) NTT Com Apr. 2000</td><td>–</td><td>–</td><td>20</td><td colspan="3">40</td><td colspan="2">70</td><td colspan="7">90</td></tr>
<tr><td>(inter-Pref.) NTT Com Mar. 2001</td><td>–</td><td>–</td><td>20</td><td colspan="3">40</td><td colspan="2">60</td><td colspan="7">80</td></tr>
<tr><td>(intra-Pref.) NTT East & West Oct. 2000</td><td>–</td><td>10</td><td>20</td><td colspan="3">30</td><td colspan="9">40</td></tr>
<tr><td>(intra-Pref.) NTT East & West Jan. 2001</td><td>–</td><td>9*</td><td>20</td><td colspan="3">30</td><td colspan="9">40</td></tr>
<tr><td>May. 2001</td><td>–</td><td>8.5</td><td>20</td><td colspan="3">30</td><td colspan="9">40</td></tr>
</table>

Shadowed columns are revised. *In January 2001 only NTT East reduced the local call rates.

[Discout System by Day of the Week and Time Zone]

Nov. 1980	• Expansion of evening discount system • Establishment of midnight discount system [• 60% discount for calls to more-than-320km zones • 9p.m.- 6a.m.]
Aug. 1981	• Establishment of Sunday/Holiday discount system [• 40% discount for Sunday/Holiday daytime calls to more-than-60km zones]
Jul. 1986	• Establishment of Saturday discount system [• 40% discount for Saturday daytime calls to more-than-60km zones]
Mar. 1990	• Expansion of midnight discount system [• 25% discount for intra-zone and short-distance calls • 45% discount for medium- and long-distance calls • 11p.m. - 6a.m.]
Mar. 1991	• Expansion of midnight discount system [• 11p.m. - 8a.m.]
Oct.1993	• Expansion of midnight discount rate [• 50 - 55% discount for medium- and long-distance calls]
Oct. 2000	• Expansion of midnight discount system [• 20% discount for calls to 20 - 60km section]

2-3-2 Mobile Phones and PHS Services

2-3-2-1 Progress of Service Provision and Movements of Carriers — Mobile Phones

December 1979	NTT Public Corp. inaugurated automobile telephone service in 23 Tokyo Metropolitan wards.
April 1987	NTT inaugurated cellular telephone service.
December 1988	Nippon Idou Tsushin Corp. (IDO) inaugurated mobile services based on the NTT large-capacity system.
July 1989	KANSAI CELLULAR TELEPHONE COMPANY inaugurated mobile services based on the TACS system.
July 1992	NTT split up its mobile communications business division, establishing NTT Mobile Communications Network, Inc. (NTT DOCOMO).
March 1993	NTT DOCOMO inaugurated mobile services based on the 800MHz band PDC system.
July 1993	NTT DOCOMO was regionally divided into 9 regional companies under the one-region-one-company system.
October 1993	NTT DOCOMO abolished the deposit money (¥100,000) system.
April 1994	The mobile terminal COAM (Customer Owned and Maintained) system was introduced. Tokyo Digital Phone Co., Ltd. and TU-KA Phone Kansai Co., Ltd. inaugurated mobile services based on the 1.5GHz band PDC system. NTT DOCOMO inaugurated mobile services based on the 1.5GHz band PDC system in Tokyo Metropolitan 23 wards.
June 1994	IDO inaugurated mobile services based on the TACS system.
January 1996	Digital TU-KA Kyushu Co., Ltd. inaugurated mobile services based on the 1.5GHz band PDC system.
December 1996	The prior notification system of mobile communications rate was started. The new subscription fee was abolished.
March 1997	NTT DOCOMO inaugurated packet communications service, "DoPa."
July 1998	DDI Cellular Group started "cdmaOne" service in Kansai, Kyushu and Okinawa.
October 1998	TU-KA Phone Kansai Co., Ltd. inaugurated prepaid cellular telephone service.
January 1999	The 11-digit numbering system was introduced to the mobile telephone service.
February 1999	NTT DOCOMO inaugurated Internet connection service, "i-mode."
March 1999	NTT DOCOMO and IDO terminated mobile services based on the NTT large-capacity system.
April 1999	DDI Cellular Group and IDO extended service areas of "cdmaOne" to cover the whole nation, and inaugurated Internet connection service, "EZweb/EZaccess."
December 1999	J-Phone Group inaugurated Internet connection service, "J-Sky."
January 2000	DDI Cellular Group and IDO inaugurated packet communications service, "PacketOne."
April 2000	DDI Cellular Group and IDO started international roaming service "GLOBAL PASSPORT".
September 2000	DDI Cellular Group and IDO terminated mobile services based on the TACS system.
October 2000	DDI, KDD and IDO merged as DDI CORPORATION (KDDI). Nine J-Phone Group companies are merged for reorganization into J-Phone East Co., Ltd., J-Phone Central Co., Ltd. and J-Phone West Co., Ltd.
November 2000	Seven companies excluding OKINAWA CELLULAR TELEPHONE of DDI Cellular Group merged as au Corp.
October 2001	KDDI merged with au.
October 2001	NTT DOCOMO started full-scale services for IMT-2000 based on the W-CDMA system.

November 2001	J-Phone Co. Ltd. as the holding company merged with J-Phone East Co., Ltd., J-Phone Central Co., Ltd. and J-Phone West Co., Ltd., and the new company was named J-Phone Co., Ltd.
November 2001	KDDI and Okinawa Cellular Telephone Company launched the cellular telephone with GPS navigation function for the first time in the Japanese market.
April 2002	KDDI and Okinawa Cellular Telephone Company started CDMA2000 1x service.
December 2002	J-Phone Co., Ltd. started 3G service using 3GPP-based W-CDMA system, and international roaming with GSM-based networks.
June 2003	NTT DOCOMO started international roaming with GSM-based networks.
October 2003	J-Phone Co., Ltd. was renamed as Vodaphone K.K.
October 2003	Vodafone inaugurated "Vodafone live!" as the 3G Internet connection service, which is also available at overseas locations.
November 2003	KDDI and Okinawa Cellular Telephone Company launched CDMA 1X WIN service.
January 2004	NTT DOCOMO inaugurated "i mode Disaster Message Board Service".
May 2004	KDDI and Okinawa Cellular Telephone launched CDMA-based international data roaming services.
July 2004	NTT DOCOMO started to provide "i-mode FeliCa" service.
October 2004	Vodafone Holdings K.K. and Vodafone K.K. were merged into new Vodafone K.K.
December 2004	Vodafone launched international video telephone roaming services.
December 2004	NTT DOCOMO launched W-CDMA type 3G mobile network services based on 3GPP, packet roaming services with GSM (GPRS) networks to make overseas i-mode connection possible, and international video telephone roaming services.
September 2005	KDDI and Okinawa Cellular Telephone Company started to provide "EZ FeliCa" service.
September 2005	KDDI and Okinawa Cellular Telephone Company started au IC card service and international roaming with GSM-based networks.
September 2005	Vodaphone started 3G data card international roaming service.
September 2005	NTT DOCOMO started to provide the "i-channel" service based on "Flash Cast".
October 2005	KDDI merged with three Tu-Ka companies.
October 2005	Vodaphone launched "Vodaphone live! NAVI", a new navigation service allowing use of network-assisted GPS function not only in Japan but also abroad.
November 2005	Vodaphone started to provide "Vodaphone live! NAVI".
November 2005	NTT DOCOMO started to provide "Push-talk" voice communication service making use of the packet network.
November 2005	KDDI and Okinawa Cellular Telephone Company started "Hello Messenger" service.
November 2005	EMOBILE Ltd. received a radio frequency license for the 1.7GHz frequency band from the Ministry of Internal Affairs and Communications and entered into mobile phone business based on the W-CDMA system.
December 2005	KDDI and Okinawa Cellular Telephone Company launched the terminal compatible with "One-Seg" ground digital telecasting service for mobile and cellular telephones.
December 2005	NTT DOCOMO started to provide a new mobile credit brand "iD".
January 2006	KDDI and Okinawa Cellular Telephone started to provide "au LISTEN MOBILE SERVICE (LISMO)".
March 2006	NTT DOCOMO launched mobile telephone terminals conforming to the one segment terrestrial digital TV service.
April 2006	NTT DOCOMO started to provide "DCMX" credit service.
April 2006	Vodafone joined the SoftBank group.
May 2006	Vodafone released a cellular phone terminal conforming to the one-segment terrestrial

	digital TV service.
August 2006	NTT DOCOMO launched "HSDPA" conforming to high-speed packet communications.
	NTT DOCOMO started to provide the "music channel" service.
September 2006	KDDI and Okinawa Cellular started "EZ Channel Plus" and "EZ News Flash" utilizing the "BCMCS".
October 2006	Vodafone changed its company name to SoftBank Mobile Corp. SoftBank Mobile started a new portal site "Yahoo! Keitai". SoftBank Mobile launched "3G high speed".
October 2006	Three cellular phone companies started a mobile number portability system.
December 2006	KDDI and Okinawa Cellular Telephone started "EV-DO Rev.A" service.
March 2007	EMOBILE started the "EM mobile broadband" HSDPA data communication service.
May 2007	NTT DOCOMO started to provide the "2in1" service, where a single mobile phone unit has the functions of two mobile phone units.
December 2007	NTT DOCOMO started to provide the "Area Mail" service.
March 2008	KDDI terminated its Tu-Ka service. KDDI and Okinawa Cellular Telephone Company started GSM-based international data-roaming service.
March 2008	EMOBILE started voice communication service based on W-CDMA, and the "EMnet" internet connection service for cellular phone terminals.
June 2008	NTT DOCOMO started to provide the "Home U" service, which allows the use of mobile phones in a broadband environment such as in the home.
July 2008	SoftBank Mobile started to provide the "Double Number" service, which allows a single mobile phone unit to manage two phone numbers and e-mail addresses.
November 2008	EMOBILE started a High-Speed Uplink Packet Access (HSUPA) data communication service.
March 2009	SoftBank Mobile started a high-speed mobile data communication service for PCs.
July 2009	EMOBILE started a High-Speed Packet Access Plus (HSPA+) data communication service.
June 2010	KDDI inaugurated ISP for smartphones "IS NET".
September 2010	NTT DOCOMO inaugurated ISP for smartphones "sp mode".
December 2010	NTT DOCOMO inaugurated LTE high-speed data communication service with maximum 75Mbps download traffic speed "Xi (crossy) service".
December 2010	EMOBILE inaugurated high-speed packet communication service with maximum 42Mbps download traffic speed "EMOBILE G4".
February 2011	SoftBank Mobile inaugurated high-speed packet communication service with maximum 42Mbps download traffic speed "ULTRA SPEED".
March 2011	NTT DOCOMO and KDDI started to provide "Disaster Message Board Service" for smartphones.
April 2011	NTT DOCOMO inaugurated SIM unlock.
May 2011	eAccess started selling EMOBILE terminals with SIM unlock.
July 2011	Inter-carrier settlement for Short Message Service (SMS) is inaugurated.
January 2012	SoftBank Mobile began providing Disaster Info.
January 2012	KDDI began providing disaster and evacuation information through its Early Warning Mail services.
January 2012	KDDI began providing mobile NFC services.
February 2012	SoftBank Mobile began providing its "SoftBank 4G" high-speed data communication service with a maximum downstream speed of 110 Mbps.
February 2012	NTT DOCOMO began delivering early warning Area Mails (tsunami warnings).
March 2012	NTT DOCOMO began providing Disaster Voice Messaging Service.

March 2012	eAccess began providing its "EMOBILE LTE" high-speed data communication service with a maximum downstream speed of 75 Mbps.
March 2012	NTT DOCOMO began selling its "Mobacas" V-High multimedia broadcasting compatible terminals (first such attempt in Japan).
March 2012	KDDI began providing tsunami warnings in its Early Warning Mail services.
March 2012	NTT DOCOMO terminated its PDC service.
April 2012	KDDI introduced the EV-DO Advanced, a technology to ease data communication congestion at wireless base stations.
June 2012	KDDI began providing a Disaster Voice Messaging Service.
July 2012	SoftBank Mobile began providing a Disaster Voice Messaging Service.
July 2012	SoftBank Mobile began providing services using the 900 MHz band.
August 2012	SoftBank Mobile began providing tsunami warnings.
August 2012	Telecommunications carriers began "all-carrier search services" for mobile phone and PHS disaster message board services and NTT EAST/WEST Disaster Message Board (web171).
September 2012	KDDI began providing the 4G LTE service based on the next-generation high-speed communication standard, LTE (Long Term Evolution).
October 2012	Business alliance between SoftBank Mobile and eAccess.
February 2013	NTT DOCOMO, China Mobile and KT developed common requirements for NFC international roaming.
February 2013	SoftBank Mobile began providing its SoftBank satellite phone service.
March 2013	eAccess began providing emergency earthquake warnings, tsunami warnings, and disaster and evacuation information through its Early Warning Mail services.
March 2013	eAccess began providing the FeliCa service.
March 2013	NTT DOCOMO, KDDI, SoftBank Mobile, and eAccess began providing mobile phone services throughout the entire Toei Subway Lines.
April 2013	NTT DOCOMO, KDDI, Okinawa Cellular, and SoftBank Mobile enabled interoperability of the Disaster Voice Messaging Service across the four mobile phone carriers.
July 2013	NTT DOCOMO, KDDI, and SoftBank Mobile began providing the LTE service at Mt. Fuji.
September 2013	SoftBank Mobile began providing international LTE roaming services.
September 2013	KDDI began providing international LTE roaming services.
October 2013	KDDI adopted the IEEE802.11ac next-generation wireless LAN standard for its au Wi-Fi SPOT public wireless LAN services.
November 2013	NTT DOCOMO, KDDI, Okinawa Cellular, SoftBank Mobile, and eAccess began using mobile phone numbers starting with 070.
November 2013	NTT DOCOMO developed a multi-band indoor base station and antenna.
January 2014	Six mobile phone and PHS carriers enabled interoperability of the Disaster Voice Messaging Service across these carriers.
March 2014	NTT DOCOMO began providing international LTE roaming services.
April 2014	NTT DOCOMO, KDDI, Okinawa Cellular, and SoftBank Mobile began delivering information on the protection of the people using the early warning Area Mails and Early Warning Mail services.
May 2014	Six mobile phone and PHS carriers standardized the number and varieties of pictographs used in text messages, including SMS, exchanged between carriers.
May 2014	KDDI introduced Carrier Aggregation, an LTE-Advanced technology based on the next-generation high-speed communication standard LTE, with a maximum receiving speed of 150 Mbps for the first time in Japan.
May 2014	NTT DOCOMO released guidelines for video distribution utilizing the next-generation

	video compression technology, HEVC.
June 2014	eAccess Ltd. and Willcom, Inc. merged.
June 2014	NTT DOCOMO developed the world's first new SIM-based authentication mini device, called Portable SIM.
June 2014	NTT DOCOMO began providing Japan's first VoLTE call service.
July 2014	eAccess Ltd. changed its company name to Ymobile Corporation.
August 2014	Ymobile started its new Y!mobile service.
October 2014	Number portability between mobile and PHS phones began.
November 2014	NTT DOCOMO started Japan's first international outbound roaming service on a TD-LTE network.
December 2014	KDDI began providing the au VoLTE next-generation voice calling service, utilizing the 4G LTE network.
December 2014	SoftBank Mobile began providing voice communication services using the VoLTE technology, a technology that enables voice communication over the LTE high-speed data communication network.
March 2015	NTT DOCOMO began providing LTE-Advanced services under the name "PREMIUM 4G" with a maximum downlink of 225 Mbps, which was the fastest in Japan.
April 2015	SoftBank Mobile Corp., SoftBank BB Corp., SoftBank Telecom Corp., and Ymobile Corporation merged together.
May 2015	The revised SIM unlocking guidelines came into effect, and NTT DOCOMO, KDDI, and SoftBank Mobile began providing SIM unlocking services based on the new guidelines.
July 2015	SoftBank Mobile Corp. changed its company name to SoftBank Corp.
October 2015	NTT DOCOMO became Japan's first telecommunications carrier to provide an international VoLTE roaming service.
March 2016	NTT DOCOMO began providing services using the world's first network function virtualization (NFV) technology that can run Evolved Packet Core (EPC) software from multiple vendors on its commercial network.
June 2016	KDDI began providing international VoLTE roaming services.
September 2016	SoftBank began providing the world's first commercial service with Massive MIMO (spatial multiplexing technology).
March 2017	NTT DOCOMO began providing communication service with a maximum downlink of 682 Mbps by introducing two new technologies: 256 QAM and 4×4 MIMO.
September 2017	KDDI began providing communication service with a maximum downlink of 708 Mbps by introducing 265 QAM and 4×4 MIMO.
May 2018	NTT DOCOMO, KDDI, and SoftBank began providing the +Message service, a new service as an extension of SMS, based on the GSMA specifications.
June 2018	NTT DOCOMO, in collaboration with China Mobile, commercialized the world's first IoT multi-vendor eSIM solution based on the GSMA 3.1 specifications.
October 2018	NTT DOCOMO, SoftBank, and KDDI each began providing services for VoLTE interconnection between different carriers.
October 2019	Rakuten Mobile began providing commercial service with the world's first end-to-end fully virtualized cloud-native network.
March 2020	NTT DoCoMo, KDDI, and SoftBank each began providing communication service using the fifth-generation mobile communication system (5G).
April 2020	Rakuten Mobile launched full-scale mobile carrier service.
September 2020	Rakuten Mobile began providing communication service using the fifth-generation mobile communication system.
October 2020	KDDI completed its succession of UQ mobile's business.
March 2021	SoftBank launched a new online-only plan under the brand name "LINEMO."

March 2021	KDDI launched a new online-only plan under the brand name “povo.”
March 2021	NTT DOCOMO launched a new online-only plan under the brand name “ahamo.”
March 2022	KDDI and Okinawa Cellular Telephone Company terminated their CDMA 1X WIN and other services for au 3G mobile phones.
May 2023	NTT DOCOMO, KDDI, Okinawa Cellular Telephone Company, SoftBank, and Rakuten Mobile introduced a “One Stop” version of mobile number portability (MNP).

Note: The transmission speeds referred to in the chronology are those at the time of the introduction of the corresponding services by the relevant companies.

2-3-2-2 Progress of Service Provision and Movements of Carriers — PHS

July 1995	DDI TOKYO POCKET TELEPHONE, Inc. DDI HOKKAIDO POCKET TELEPHONE Inc., NTT Central Personal Communications Network Inc., and NTT Hokkaido Personal Communications Network Inc. inaugurated services. After October 1995, 7 companies of DDI POCKET TELEPHONE Group, 7 companies of NTT Personal Communications Network Group and 10 companies of ASTEL Group inaugurated services.
February 1997	The new subscription fee was abolished.
December 1998	Nine companies of NTT Personal Communications Network Group assigned their business to nine companies of NTT DOCOMO Group.
April 1999	ASTEL Tokyo Corporation was merged into Tokyo Telecommunication Network Co., Inc. NTT DOCOMO Group inaugurated 64kbps data communications service.
November 1999	ASTEL Hokkaido Corporation assigned its business to HOKKAIDO TELECOMMUNICATION NETWORK CO., Inc.
January 2000	Nine companies of DDI POCKET TELEPHONE Group were amalgamated as DDI POCKET Inc.
September 2000	ASTEL Tohoku Corporation assigned its business to Tohoku Intelligent Telecommunication Co., Inc.
November 2000	ASTEL Chubu and CHUBU TELECOMMUNICATIONS CO., INC. merged. ASTEL KANSAI CORPORATION assigned its business to K-Opticom Corporation.
April 2001	ASTEL KYUSHU assigned its business to Kyushu Telecommunication Network Co., Inc.
August 2001	DDI Pocket Inc. inaugurated fixed-rate data communication service.
October 2001	Astel Chugoku Corporation assigned its business to Chugoku Information System Service Co., Inc.
December 2001	Astel Hokuriku Corporation assigned its business to Hokuriku Telecommunication Network Co., Inc.
March 2002	Astel Shikoku Corporation assigned its business to Shikoku Information and Telecommunication Network Company, Incorporated.
April 2002	Shikoku Information and Telecommunication Network Company, Incorporated changed the company name to STNet Incorporated.
August 2002	Tokyo Telecommunication Network Company, Incorporated assigned its PHS business to Magic Mail Inc.
October 2002	Magic Mail Inc. was merged with Yozan Inc.
April 2003	NTT DOCOMO group inaugurated fixed-rate data communication service.
July 2003	Chugoku Telecommunication Network merged with Chugoku Information System Service and reorganized as Energia Communications.
November 2003	Kyushu Telecommunications Network Co., Inc. terminated their PHS telephone service.
March 2004	Hokkaido Telecommunications Network Co., Inc. terminated their PHS telephone service.
May 2004	Hokuriku Telecommunications Network Co., Inc. terminated their PHS telephone service.
September 2004	K-Opticom Corporation terminated the PHS voice telephone service out of their PHS services.
October 2004	DDI Pocket, Inc. became independent from the KDDI group.
December 2004	Energia Communications ceased to provide PHS voice telephone service out of their PHS services.
January 2005	Astel Okinawa transferred goodwill to WILLCOM Okinawa.

February 2005	DDI Pocket, Inc. changed the name to WILLCOM, Inc.
May 2005	STNet ceased to provide their PHS telephone service.
May 2005	Chubu Telecommunications Co., Inc. ceased to provide their PHS communication service.
May 2005	Willcom started "Willcom Teigaku Plan" fixed-rate service.
June 2006	YOZAN terminated its PHS telephone service.
December 2006	Tohoku Intelligent Telecommunication terminated its PHS telephone service.
October 2007	Energia Communications terminated PHS services.
January 2008	NTT DOCOMO Group terminated their PHS services.
December 2010	Willcom started the “Fixed Rate with Anyone” service.
September 2011	K-Opticom terminated its PHS service.
June 2014	Willcom merged with eAccess (eAccess Ltd.).
January 2021	SoftBank terminated its PHS service.

2-3-3 International Telephone Services

2-3-3-1 Progress of Service Provision and Movements of Carriers

- In October 1989, International Telecom Japan Inc. (ITJ) and International Digital Communications Inc. (IDC) introduced services with 23% lower rates than those of Kokusai Denshin Denwa Co.,Ltd. (KDD)
- From 1989 through 1996 KDD implemented rate reductions eight times, and ITJ and IDC five times, resulting in a steady shift toward less expensive rates.

October 1998	DDI Corporation (DDI) started international telephone services with the level of charge set at ¥240 for a daytime 3-minute call to U.S. MCI Worldcom Japan, Inc. (WCOM) started international telephone services with the level of charge set at ¥248 for a daytime 3-minute call to U.S.
December 1998	KDD reduced charges for calls to all destinations (230 countries and areas). The average reduction rate was about 10.6%. As the result, a daytime 3-minute call to U.S. cost ¥240. Japan Telecom (JT) reduced charges for calls to 28 destinations. The average reduction rate was about 8.6%. A daytime 3-minute call to U.S. cost ¥240. IDC reduced charges for calls to 23 destinations. The average reduction rate was about 9.0%. A daytime 3-minute call to U.S. cost ¥240. WCOM reduced charges. A daytime 3-minute call to U.S. cost ¥150.
January 1999	DDI reduced charges for calls to 25 destinations. The average reduction rate was about 8.4%. A daytime 3-minute call to U.S. cost ¥168. JT reduced charges for calls to 97 destinations. The average reduction rate was about 2.2%. IDC reduced charges for calls to 51 destinations. The average reduction rate was about 3.5%.
March 1999	DDI reduced charges for calls to 27 destinations, with a main target of reduction on calls placed during 23:00 to 08:00 of the following day. The average reduction rate was about 5.8%.
July 1999	Tokyo Telecommunication Network Co.,Inc. (TTNet) started international telephone services with the level of charge set at ¥168 for a daytime 3-minute call to U.S.
October 1999	JT reduced charges for all destinations (223 countries and areas). The average reduction rate was about 10.3%. A daytime 3-minute call to U.S. cost ¥180. Cable & Wireless IDC reduced charges for calls to 192 destinations. The average reduction rate was about 10.9%. A daytime 3-minute call to U.S. cost ¥180. NTT Communications Corp. started international telephone services with the level of charge set at ¥180 for a daytime 3-minute call to U.S.
November 1999	KDD reduced charges for calls to all destinations (231 countries and areas). The average reduction rate was about 11.1%. A daytime 3-minute call to U.S. cost ¥180. DDI reduced charges for calls to 38 destinations. The average reduction rate was about 8.4%. A daytime 3-minute call to U.S. cost ¥156. TTNet reduced charges for calls to 58 destinations. The average reduction rate was about 11%. A daytime 3-minute call to U.S. cost ¥132.
December 1999	KDD reduced charges for cellular/PHS-originated calls to all destinations (231 countries/areas). The average reduction rate was about 11.9%.
February 2000	KDD reduced charges for calls to 17 destinations (Taiwan, China, U.K., France, Germany, etc.). The average reduction rate was about 1.4%.
October 2000	DDI, KDD and IDO were merged as KDDI.
April 2001	Fusion Communications started international telephone services, establishing the all-time flat rate system. The charge for 3-minute calls to U.S. is ¥90.
September 2001	Fusion Communications Corporation reduced the charges for calls to all destinations (230 countries and areas). A three-minute call to the U.S. cost ¥45.

April 2003	POWEREDCOM merged with TTNet, and the new company was named POWEREDCOM, Inc.
July 2004	The telephone business of POWEREDCOM is merged with FUSION COMMUNICATIONS CORP.
October 2006	Japan Telecom Co. Ltd. changed its company name to SoftBank Telecom Corp.
April 2015	SoftBank Mobile Corp., SoftBank BB Corp., SoftBank Telecom Corp., and Ymobile Corporation merged together to form SoftBank Mobile Corp.
July 2015	SoftBank Mobile Corp. changed its company name to SoftBank Corp.
December 2015	Fusion Communications Corp. changed its company name to Rakuten Communications Corp.
July 2019	Rakuten Communications Corp. transferred its international telephone service to Rakuten Mobile, Inc. through a company split.

2-3-4 Leased Circuit and Data Transmission Services

2-3-4-1 Progress of Service Provision and Movements of Carriers

• Progress of Leased Circuit Service Provision

(NTT)

December 1997	NTT started "Digital Access 128" as short-distance economy service.
April 1998	NTT started "Digital Access 1500" service.
August 1998	NTT started "Digital Reach" as medium- and long-distance economy service.
December 1998	NTT started "ATM SHARE LINK" as partial band assurance type exclusively for ATM.
October 1999	NTT Communications started "Gigaway" service.
March 2000	NTT Communications started "Air Access" service.
April 2001	NTT East and West started "Digital Access 6000" service.
November 2001	NTT East started "Metro High Link" service.
June 2002	NTT East started "Super-high Link" service.
July 2002	NTT West started "Giga Data Link" service.
October 2002	NTT Communications started "EtherArcstream" service.
June 2004	NTT Communications started “GIGASTREAM” service.
December 2008	NTT Communications started “GIGASTREAM Premium Ether” service.
May 2011	NTT Communications started to provide “Arcstar Universal One”.

(Long-Distance and International Carriers)

April 1998	KDDI (TWJ) started to provide leased circuit service for ATM.
October 1998	Long-distance and International NCCs started economy services.
September to October 1999	Long-distance and International NCCs acquired rate setting right and started end-to-end rate services.
January 2000	Global Access started domestic and international leased circuit service.
July 2000	Japan Telecom started domestic wide-band leased circuit service.
October 2002	Japan Telecom started international wide-band leased circuit service.

(Regional Carriers)

April 1997	Nine electric power companies started joint high-speed digital transmission service.
January 1998	TTNet started FDDI leased circuit service.
April 1998	TTNet started leased circuit service for ATM.
May 1998	Ten electric power companies completed nationwide linkage of high-speed digital transmission services.
October 1998	Nine electric power companies started linkage of ATM leased circuit services.
August 1999	Ten electric power companies completed nationwide linkage of economy services.
April 2001	TTNet started to provide "PeneLink (leased circuit)" (Ethernet leased circuit service).
September 2001	Keio Network Communications started to provide "Express-Ether" service.
April 2002	Osaka Media Port started Ether leased circuit service.
June 2002	Chubu Telecommunication started optical fiber leased circuit service.
April 2003	Osaka Media Port started Ether Network service (W-Link).

(Regional CATV)

April 2002	Katch Network started optical fiber leased circuit service.
December 2002	Himawari Network started optical fiber leased circuit service.
December 2002	My Television started regional LAN services.

• Progress of Data Transmission Service Provision

(NTT)

December 1996	NTT started OCN service.
August 1999	NTT Communications started to provide OBN (Open Business Network) service.
September 1999	NTT Communications started to provide "Arcstar Value Access" service.
May 2000	NTT East and West started to provide Wide LAN Service.
July 2000	NTT Communications started "Super VPN (current Arcstar IP-VPN)" service.
July 2000	NTT DOCOMO and NTT Communications jointly started to provide "RALS (Remote Access Line Service)".
September 2000	NTT East started to provide FLET's Office".
October 2000	NTT Communications started to provide "Broadband Access" service.
October 2000	NTT East and West started to provide "Mega Data Nets" service.
December 2000	NTT Communications started to provide "Giga Ether Platform" service.
January 2001	NTT Communications started to provide "Arcstar Global IP-VPN" service.
March 2001	NTT East started to provide "Metro Ether" service.
April 2001	NTT Communications started to provide "e-VLAN" service.
May 2001	NTT West started to provide "Urban Ether" service.
March 2002	NTT East started to provide "FLET's Group Access" service.
March 2002	NTT East started to provide "Super Wide LAN Service".
March 2002	NTT West started to provide "Wide LAN Plus" service.
March 2003	NTT East started to provide "FLET's Office Wide" service.
April 2003	NTT Communications started to provide "Super HUB" service.
May 2003	NTT Communications started to provide "FLEXGIGAWAY" service.
July 2003	NTT East started to provide "Flat Ether" service.
October 2003	NTT West started to provide "Flat Ether" service.
December 2003	NTT East started to provide the Smart Ether service.
June 2004	NTT Communications started to provide the “Group-VPN” service.
April 2006	NTT West started to provide the “Business Ether” service.
May 2006	NTT East started to provide the “Business Ether” service.
July 2009	NTT Communications started to provide the “Group-Ether” service.
May 2011	NTT Communications started to provide “Arcstar Universal One”.

(Long-Distance and International Carriers)

April 1997	Long-distance and International NCCs sequentially started to provide computer network services.
April 1999	Japan Telecom started to provide international cell relay service.
April 2000	Japan Telecom started to provide Solteria (IP-VPN) service.
October 2000	KDDI started to provide ANDROMEGA IP-VPN service.
February 2001	Fusion Communications started to provide FUSION IP-VPN service.
October 2001	Japan Telecom started to provide "Wide-Ether" (wide-area LAN).
December 2001	Cable & Wireless IDC started to provide "High-speed Ethernet Service".

December 2001 KDDI started to provide "Ether-VPN" service.

September 2002 Cable & Wireless IDC started to provide "IP-VPN QoS" service.

November 2002 Japan Telecom started to provide "ASSOCIO (MLPS Traffic Switching Service)".

August 2012 SoftBank Telecom began providing its White Cloud SmartVPN service.

(Regional Carriers)

From September 1997 Power company based NCCs sequentially started to provide computer network services.

March 2001 Hokkaido Telecommunication Network, Inc started to provide wide-area Ethernet service "L2L".

April 2001 Poweredcom started to provide "Powered Ethernet" wide-area Ethernet connection service.

April 2001 TTNet started to provide "Pene-Link (Multi-access)" (wide-area Ethernet connection service).

June 2001 K-Opticom started to provide IP-VPN service.

July 2001 Poweredcom started to provide "Powered-IP MPLS" (IP-VPN connection service).

August 2001 Chugoku Telecommunication Network started to provide Ethernet communication network service "V-LAN".

June 2002 Keio Network Communications started to provide "Multi-Express Ether" service.

July 2003 Chugoku Telecommunication Network merged with Chugoku Information System Service and reorganized as Energia Communications.

January 2003 Chubu Telecommunication started to provide band-assured type Ether network service "CTC Ether Link".

June 2005 Chubu Telecommunication started to provide "CTC Ether DIVE" wide-area Ethernet service.

(Regional CATV)

December 1995 Himawari Network started to provide cell relay service.

November 1997 Katch Network started to provide cell relay service.

April 1998 MICS Network started to provide ATM switching service.

September 1999 MICS Network started to provide wide-area LAN service.

Chapter 3
Situation of TCA Members

Member List

***Symbols representing types of provided services**

1. Subscriber telephone
2. ISDN (excluding switched telephone, public telephone and international digital communications service)
3. Switched telephone (excluding international telephone)
4-①. International telephone
4-②. International ISDN
5. Public telephone
6-①. Mobile phone (using a 3.9-generation mobile communication system)
6-②. Mobile phone (using a 5th generation mobile communication system)
6-③. Mobile phone (not using 3.9–4th generation mobile communication systems or a 5th generation mobile communication system)
7. PHS
8-①. IP phone (using 050/0AB-J number)
8-②. IP phone (not using 050/0AB-J number)
9. Wireless landline telephone
10. Satellite mobile communication service
11. FMC service
12. Internet connection service
13-①. FTTH access service
13-②. FTTH access service (using VDSL or other facility)
14. DSL access service
15. FWA access service
16. CATV access service
17. Mobile/PHS access service
18. 3.9-generation mobile phone access service
19. Fifth generation mobile communication access service
20. Local 5G service
21. Frame relay service
22. ATM switch service
23. Public wireless LAN access service
24-①. BWA service (nationwide BWA service)
24-②. BWA service (regional BWA service)
24-③. BWA service (private BWA service)
25. IP-VPN service
26. Wide-area Ethernet service
27. Satellite access service
28-①. Leased line service (domestic)
28-②. Leased line service (international)
29. Unlicensed LPWA service
30. Value added service using telecommunication services in 1 to 28 above
31. Internet-related service (excluding IP phone)
32-①. Mobile virtual network service (for mobile phones)
32-②. Mobile virtual network service (for PHS)
32-③. Mobile virtual network service (for local 5G service)
32-④. Mobile virtual network service (for BWA service)
33-①. Telecommunications service for domain name facilities (as specified in Article 59-2, paragraph (1), item (i)-(a))
33-②. Telecommunications service for domain name facilities (as specified in Article 59-2, paragraph (1), item (i)-(b))
33-③. Telecommunications service for domain name facilities (as specified in Article 59-2, paragraph (1), item (ii))
34-①. Telegram (involving acceptance and delivery jobs)
34-②. Telegram (without acceptance and delivery jobs)
35. Telecommunication services other than 1 through 33 above

* Representation of service areas

Hokkaido

Tohoku District: Aomori, Iwate, Miyagi, Akita, Yamagata, and Fukushima

Kanto District: Ibaraki, Tochigi, Gunma, Saitama, Tokyo, Chiba, Kanagawa, and Yamanashi

Shin-etsu District: Nagano and Niigata

Hokuriku District: Toyama, Ishikawa, and Fukui

Tokai District: Shizuoka, Aichi, Gifu, and Mie

Kinki District: Shiga, Kyoto, Osaka, Hyogo, Nara, and Wakayama

Chugoku District: Tottori, Shimane, Okayama, Hiroshima, and Yamaguchi

Shikoku District: Tokushima, Kagawa, Ehime, and Kochi

Kyushu District: Fukuoka, Saga, Nagasaki, Kumamoto, Oita, Miyazaki, and Kagoshima

Okinawa

*Compiled from the data in the questionnaire survey covering member companies based on the "Registered Member Companies" (Ministry of Internal Affairs and Communications).

As of July 1, 2023

Company	Service areas	Types of services																																														
		1	2	3	4①	4②	5	6①	6②	6③	7	8①	8②	9	10	11	12	13①	13②	14	15	16	17	18	19	20	21	22	23	24①	24②	24③	25	26	27	28①	28②	29	30	31	32①	32②	32③	32④	33①	33②	33③	34①	34②	35
NIPPON TELEGRAPH AND TELEPHONE CORPORATION																																																		
NIPPON TELEGRAPH AND TELEPHONE EAST CORPORATION	Hokkaido, Tohoku District, Kanto District, and Shin-etsu District	○	○				○					○	○					○	○	○		○				○		○						○		○			○	○								○		○
NIPPON TELEGRAPH AND TELEPHONE WEST CORPORATION	Hokuriku District, Tokai District, Kinki District, Chugoku District, Shikoku District, Kyushu District, and Okinawa	○	○				○					○	○					○	○	○	○	○						○	○					○		○			○	○								○		○
KDDI CORPORATION	Nationwide			○	○	○		○	○			○			○		○	○	○				○	○	○				○				○	○	○	○	○		○	○	○			○				○		○
SoftBank Corp.	Nationwide	○	○	○	○	○		○	○	○		○	○		○		○	○	○	○			○	○	○				○				○	○	○	○	○		○	○	○			○					○	○
ARTERIA Networks Corporation	Nationwide			○								○	○				○	○	○														○	○		○	○		○	○	○	○		○						
NTT Communications Corporation	Nationwide			○	○	○						○	○		○		○	○	○										○				○	○		○	○		○	○	○								○	○
East Japan Railway Company	Nationwide																												○				○	○										○						○
SKY Perfect JSAT Corporation	Nationwide											○			○		○																	○	○	○	○		○											○

Company	Service areas	Types of services																																														
		1	2	3	4①	4②	5	6①	6②	6③	7	8①	8②	9	10	11	12	13①	13②	14	15	16	17	18	19	20	21	22	23	24①	24②	24③	25	26	27	28①	28②	29	30	31	32①	32②	32③	32④	33①	33②	33③	34①	34②	35
PCCW Global (Japan) K.K.	Nationwide																○																				○													
Sony Network Communications Inc.	Nationwide											○	○				○	○	○														○	○		○		○	○	○	○			○						
Hokkaido Telecommunication Network Co., Inc	Hokkaido																○																	○		○				○										
Hokkaido Telecommunication Network Co., Inc	Tokyo																																							○										
TOHKnet Co., Inc.	Tohoku District, and Niigata				○							○					○												○				○	○		○			○	○	○		○	○						
Hokuriku Telecommunication Network Co.,Inc	Hokuriku District											○					○																○	○		○	○		○	○										
Chubu Telecommunications Company, Incorporated	Tokai District, and Nagano											○	○		○		○	○	○														○	○	○	○				○	○	○		○						○
OPTAGE Inc.	Nationwide				○							○					○	○	○														○	○	○	○			○	○	○			○						○
Enecom,Inc.	Chugoku District											○	○				○	○	○															○		○			○	○	○									
STNet, Incorporated	Nationwide											○					○	○	○														○	○		○			○		○			○						○
QTnet, Inc.	Nationwide											○					○	○	○							○			○				○	○		○	○	○		○	○			○						
OTNet Company, Incorporated	Okinawa											○					○	○	○														○	○		○		○												
Japan Digital Serve Corporation	Kanto District																																		○														
J-POWER Telecommunication Service Co., Ltd.	Nationwide											○	○		○	○	○																○	○	○	○				○	○									
LCV Corporation	Shinetsu District											○					○	○	○			○							○		○					○					○									
Kintetsu Cable Network Co., Ltd.	Nara, Osaka, Kyoto, Mie, and Aichi											○					○	○	○			○									○		○			○				○										
Kintetsu Cable Network Co., Ltd.	Nationwide																																								○			○						
its communications Inc.	Kanto District (Tokyo and Kanagawa)											○					○	○	○	○		○							○		○					○				○	○									
Cable Television Shinagawa inc.	Tokyo											○					○	○	○			○											○							○	○									
Newmedia Co., Ltd.	Yonezawa City, Nan'yo City, Takahata Town, and Kawanishi Town in Yamagata Hakodate City, Hokuto City, and Nanae Town in Hokkaido Niigata City in Niigata Fukushima City in Fukushima																○	○	○			○							○		○					○					○									
CTY.co.,Ltd	Mie											○					○	○											○		○					○				○	○									

Company	Service areas	Types of services																																														
		1	2	3	4 ①	4 ②	5	6 ①	6 ②	6 ③	7	8 ①	8 ②	9	10	11	12	13 ①	13 ②	14	15	16	17	18	19	20	21	22	23	24 ①	24 ②	24 ③	25	26	27	28 ①	28 ②	29	30	31	32 ①	32 ②	32 ③	32 ④	33 ①	33 ②	33 ③	34 ①	34 ②	35
TOKYO CABLE NETWORK, INC.	Hokkaido, Iwate, Miyagi, Yamagata, Tochigi, Yamanashi, Tokyo, Kanagawa, Nagano, Nigata, Toyama, Ishikawa, Fukui, Shizuoka, Gifu, Mie, Nara, Wakayama, Kyoto, Hyogo, Tottori, Shimane, Okayama, Hiroshima, Tokushima, Kagawa, Ehime, Fukuoka, Nagasaki, Kumamoto, Oita, Miyazaki, Kagoshima, Okinawa											○					○	○	○		○	○							○		○			○		○			○	○				○						
JCOM Co., Ltd.	Nationwide																○																			○			○	○	○			○						○
MICS NETWORK CORPORATION	Aichi	○																○	○			○							○		○	○	○																	
advanscope inc.	Tokai District							○									○	○				○							○		○																			
TOKAI Communications Corporation	Nationwide											○					○	○	○	○								○					○	○		○	○		○	○	○			○						○
Cable Networks AKITA	Tohoku District											○					○	○				○				○			○		○		○			○				○	○			○						
Matsusaka CATV Station Co.,Ltd. (MCTV)	Mie																○	○				○									○					○					○									
COMMUNITY NETWORK CENTER INCORPORATED	Tokai District																												○		○			○					○		○									○
Igaueno Cable Television Co.,Ltd.	Mie											○					○	○	○			○							○		○			○		○														
Ichihara Cable Television Corporation	Ichihara City, and Midori Ward, Chiba City											○					○	○	○		○	○									○					○			○	○	○									
CHUKAI CABLE TELEVISION SYSTEM OPERATOR	Western Tottori											○					○	○				○									○					○				○	○									
IRUMA CABLE TELEVISION CO., LTD	Kanto District											○						○				○									○					○			○	○	○									
NTT DOCOMO, INC.	Nationwide			○	○			○	○	○		○			○		○	○	○				○	○	○				○				○	○	○	○			○	○	○			○						
OKINAWA CELLULAR TELEPHONE COMPANY	Okinawa				○			○	○			○			○		○	○	○				○	○	○				○						○	○			○	○	○			○						
Rakuten Mobile, Inc.	Nationwide		○	○				○	○			○	○				○	○	○	○			○	○	○															○	○			○						○
Tokyo Telemessage Inc.	Nationwide																																																	○
AVICOM JAPAN CO., LTD.	Nationwide																																			○														

Company	Service areas	Types of services																																														
		1	2	3	4①	4②	5	6①	6②	6③	7	8①	8②	9	10	11	12	13①	13②	14	15	16	17	18	19	20	21	22	23	24①	24②	24③	25	26	27	28①	28②	29	30	31	32①	32②	32③	32④	33①	33②	33③	34①	34②	35
Kansai Airports Technical Services Co.,Ltd	Osaka																○																							○										
UQ Communications Inc.	Nationwide																○													○				○					○		○									○

Telecom Data Book 2023
(Compiled by TCA)

Planned / Edited / Published by Telecommunications Carriers Association

Koshin Bldg. 2F
1-10 Kanda-Ogawamachi, Chiyoda-ku, Tokyo 101-0052, Japan
Tel. 03-5577-5845 Fax. 03-5296-5520
https://www.tca.or.jp/

Co-edited / Printed HARIU Communications Co., Ltd.

Published: February 2024

Price: ¥2,200 (¥2,000 + 10% tax)
 ISBN978-4-906932-23-8